CAMBRIDGE LIBRARY COLLECTION

Books of enduring scholarly value

Life Sciences

Until the nineteenth century, the various subjects now known as the life sciences were regarded either as arcane studies which had little impact on ordinary daily life, or as a genteel hobby for the leisured classes. The increasing academic rigour and systematisation brought to the study of botany, zoology and other disciplines, and their adoption in university curricula, are reflected in the books reissued in this series.

Problems of Genetics

A key figure in the field of evolutionary biology, William Bateson (1861–1926) revived Mendelian methods of analysis to develop Darwin's theory of evolution, thereby pioneering the study of genetics. In these lectures, published at Yale in 1913, Bateson systematically chronicles the era's conflicting and developing theories on taxonomy, speciation, variation and hybridisation, and includes his own thoughts on continuous and discontinuous variation and its causes. Drawing on the comparative physiology and anatomy of species that he knew from his wide experience, citing detailed examples from across the taxonomic kingdoms, Bateson brings to life this exciting time in biology. Because the theories central to the modern understanding of genetics, heredity and evolution were formed at this time, this work remains valuable and relevant to students of biology and the history of science.

Cambridge University Press has long been a pioneer in the reissuing of out-of-print titles from its own backlist, producing digital reprints of books that are still sought after by scholars and students but could not be reprinted economically using traditional technology. The Cambridge Library Collection extends this activity to a wider range of books which are still of importance to researchers and professionals, either for the source material they contain, or as landmarks in the history of their academic discipline.

Drawing from the world-renowned collections in the Cambridge University Library and other partner libraries, and guided by the advice of experts in each subject area, Cambridge University Press is using state-of-the-art scanning machines in its own Printing House to capture the content of each book selected for inclusion. The files are processed to give a consistently clear, crisp image, and the books finished to the high quality standard for which the Press is recognised around the world. The latest print-on-demand technology ensures that the books will remain available indefinitely, and that orders for single or multiple copies can quickly be supplied.

The Cambridge Library Collection brings back to life books of enduring scholarly value (including out-of-copyright works originally issued by other publishers) across a wide range of disciplines in the humanities and social sciences and in science and technology.

Problems of Genetics

William Bateson

CAMBRIDGE UNIVERSITY PRESS

Cambridge, New York, Melbourne, Madrid, Cape Town,
Singapore, São Paolo, Delhi, Mexico City

Published in the United States of America by Cambridge University Press, New York

www.cambridge.org
Information on this title: www.cambridge.org/9781108053082

© in this compilation Cambridge University Press 2012

This edition first published 1913
This digitally printed version 2012

ISBN 978-1-108-05308-2 Paperback

YALE UNIVERSITY

MRS. HEPSA ELY SILLIMAN MEMORIAL LECTURES

PROBLEMS OF GENETICS

The right hand figure shows *Colaptes cafer*, the left *Colaptes auratus*.

PROBLEMS OF GENETICS

BY

WILLIAM BATESON, M.A., F.R.S.

DIRECTOR OF THE JOHN INNES HORTICULTURAL INSTITUTION, HON. FELLOW OF ST. JOHN'S
COLLEGE, CAMBRIDGE, AND FORMERLY PROFESSOR OF BIOLOGY IN THE UNIVERSITY

WITH ILLUSTRATIONS

New Haven: Yale University Press
London: Humphrey Milford
Oxford University Press
MCMXIII

THE SILLIMAN FOUNDATION

In the year 1883 a legacy of about eighty-five thousand dollars was left to the President and Fellows of Yale College in the city of New Haven, to be held in trust, as a gift from her children, in memory of their beloved and honored mother, Mrs. Hepsa Ely Silliman.

On this foundation Yale College was requested and directed to establish an annual course of lectures designed to illustrate the presence and providence, the wisdom and goodness of God, as manifested in the natural and moral world. These were to be designated as the Mrs. Hepsa Ely Silliman Memorial Lectures. It was the belief of the testator that any orderly presentation of the facts of nature or history contributed to the end of this foundation more effectively than any attempt to emphasize the elements of doctrine or of creed; and he therefore provided that lectures on dogmatic or polemical theology should be excluded from the scope of this foundation, and that the subjects should be selected rather from the domains of natural science and history, giving special prominence to astronomy, chemistry, geology, and anatomy.

It was further directed that each annual course should be made the basis of a volume to form part of a series constituting a memorial to Mrs. Silliman. The memorial fund came into the possession of the Corporation of Yale University in the year 1901; and the present volume constitutes the fifth of the series of memorial lectures.

PREFACE

This book gives the substance of a series of lectures delivered in Yale University, where I had the privilege of holding the office of Silliman Lecturer in 1907.

The delay in publication was brought about by a variety of causes.

Inasmuch as the purpose of the lectures is to discuss some of the wider problems of biology in the light of knowledge acquired by Mendelian methods of analysis, it was essential that a fairly full account of the conclusions established by them should first be undertaken and I therefore postponed the present work till a book on Mendel's Principles had been completed.

On attempting a more general discussion of the bearing of the phenomena on the theory of Evolution, I found myself continually hindered by the consciousness that such treatment is premature, and by doubt whether it were not better that the debate should for the present stand indefinitely adjourned. That species have come into existence by an evolutionary process no one seriously doubts; but few who are familiar with the facts that genetic research has revealed are now inclined to speculate as to the manner by which the process has been accomplished. Our knowledge of the nature and properties of living things is far too meagre to justify any such attempts. Suggestions of course can be made: though, however, these ideas may have a stimulating value in the lecture room, they look weak and thin when set out in print. The work which may one day give them a body has yet to be done.

The development of negations is always an ungrateful task apt to be postponed for the positive business of experiment. Such work is happily now going forward in most of the centers of scientific life. Of many of the subjects here treated we already know more than we did in 1907. The delay in production has made it possible to incorporate these new contributions. The book makes no pretence at being a treatise and the number of illustrative cases has been kept within a moderate compass. A good many of the examples have been chosen from American natural history, as being appropriate to a book intended primarily for American readers. The facts are largely given on the authority of others, and I wish to express my gratitude for the abundant assistance received from American colleagues, especially from the staffs of the American Museum in New York, and of the Boston Museum of Natural History. In connexion with the particular subjects personal acknowledgments are made.

Dr. F. M. Chapman was so good as to supervise the preparation of the coloured Plate of *Colaptes*, and to authorize the loan of the Plate representing the various forms of *Helminthophila*, which is taken from his *North American Warblers*.

I am under obligation to Messrs. Macmillan & Co., for permission to reproduce several figures from *Materials for the Study of Variation*, illustrating subjects which I wished to treat in new associations, and to M. Leduc for leave to use Fig. 9.

In conclusion I thank my friends in Yale for the high honour they did me by their invitation to contribute to the series of Silliman Lectures, and for much kindness received during a delightful sojourn in that genial home of learning.

TABLES OF CONTENTS.

TABLES OF CONTENTS

PROBLEMS OF GENETICS

PROBLEMS OF GENETICS

CHAPTER I

INTRODUCTORY

THE purpose of these lectures is to discuss some of the familiar phenomena of biology in the light of modern discoveries. In the last decade of the nineteenth century many of us perceived that if any serious advance was to be made with the group of problems generally spoken of as the Theory of Evolution, methods of investigation must be devised and applied of a kind more direct and more penetrating than those which after the general acceptance of the Darwinian views had been deemed adequate. Such methods obviously were to be found in a critical and exhaustive study of the facts of variation and heredity, upon which all conceptions of evolution are based. To construct a true synthetic theory of Evolution it was necessary that variation and heredity instead of being merely postulated as axioms should be minutely examined as phenomena. Such a study Darwin himself had indeed tentatively begun, but work of a more thorough and comprehensive quality was required. In the conventional view which the orthodoxy of the day prescribed, the terms variation and heredity stood for processes so vague and indefinite that no analytical investigation of them could be contemplated. So soon, however, as systematic inquiry into the natural facts was begun it was at once found that the accepted ideas of variation were unfounded. Variation was seen very frequently to be a definite and specific phenomenon, affecting different forms of life in different ways, but in all its diversity showing manifold and often obvious indications of regularity. This observation was not in its essence novel. Several examples of definite variation had been well known to

2   I

Darwin and others, but many, especially Darwin himself in his later years, had nevertheless been disposed to depreciate the significance of such facts. They consequently then lapsed into general disparagement. Upon more careful inquiry the abundance of such phenomena proved to be far greater than was currently supposed, and a discussion of their nature brought into prominence a consideration of greater weight, namely that the differences by which these definite or discontinuous variations are constituted again and again approximate to and are comparable with the class of differences by which species are distinguished from each other.

The interest of such observations could no longer be denied. The more they were examined the more apparent it became that by means of the facts of variation a new light was obtained on the physiological composition and capabilities of living things. Genetics thus cease to be merely a method of investigating theories of evolution or of the origin of species but provide a novel and hitherto untried instrument by which the nature of the living organism may be explored. Just as in the study of non-living matter science began by regarding the external properties of weight, opacity, colour, hardness, mode of occurrence, etc., noting only such evidences of chemical attributes and powers as chance spontaneously revealed; and much later proceeded to the discovery that these casual manifestations of chemical properties, rightly interpreted, afford a key to the intrinsic nature of the diversity of matter, so in biology, having examined those features of living things which ordinary observations can perceive, we come at last to realize that when studied for their own sake the properties of living organisms in respect of heredity and variation are indications of their inner nature and provide evidences of that nature which can be obtained from no other source.

While such ideas were gradually forming in our minds, came the rediscovery of Mendel's work. Investigations which before had only been imagined as desirable now became easy to pursue, and questions as to the genetic inter-relations and compositions of varieties can now be definitely answered. Without prejudice to what the future may disclose whether by way of limitation or extension of Mendelian method, it can be declared with confidence and certainty that we have now the means of beginning an analysis of living organisms, and distinguishing many of the units or factors which essentially determine and cause the development of their several attributes.

Briefly put, the essence of Mendelism lies in the discovery of the existence of unit characters or factors. For an account of the Mendelian method, how it is applied and what it has already accomplished, reference must be made to other works.[1] With this part of the subject I shall assume a sufficient acquaintance. In these lectures I have rather set myself the task of considering how certain problems appear when viewed from the standpoint to which the application of these methods has led us. It is indeed somewhat premature to discuss such questions. The work of Mendelian analysis is progressing with great rapidity and anything I can say may very soon be superseded as out of date. Nevertheless a discussion of this kind may be of at least temporary service in directing inquiry to the points of special interest.

THE PROBLEM OF SPECIES AND VARIETY

Nowhere does our new knowledge of heredity and variation apply more directly than to the problem what is a species and what is a variety? I cannot assert that we are already in a position to answer this important question, but as will presently appear, our mode of attack and the answers we expect to receive are not those that were contemplated by our predecessors. If we glance at the history of the scientific conception of Species we find many signs that it was not till comparatively recent times that the definiteness of species became a strict canon of the scientific faith and that attempts were made to give precise limits to that conception. When the diversity of living things began to be accurately studied in the sixteenth and seventeenth

[1] In *Mendel's Principles of Heredity* (Cambridge University Press, 1909) I have dealt with this subject, giving an account of the principal facts discovered up to the beginning of 1909.

centuries names were applied in the loosest fashion, and in giving a name to an animal or a plant the naturalists of those times had no ulterior intention. Names were bestowed on those creatures about which the writer proposed to speak. When Gesner or Aldrovandi refer to all the kinds of horses, unicorns, dogs, mermaids, etc., which they had seen or read of, giving to each a descriptive name, they do not mean to "elevate" each named kind to "specific rank"; and if anyone had asked them what they meant by a species, it is practically certain that they would have had not the slightest idea what the question might imply, or any suspicion that it raised a fundamental problem of nature.

Spontaneous generation being a matter of daily observation, then unquestioned, and supernatural events of all kinds being commonly reported by many witnesses, transmutation of species had no inherent improbability. Matthioli,[2] for instance, did not expect to be charged with heresy when he declared *Stirpium mutatio* to be of ordinary occurrence. After giving instances of induced modifications he wrote, "Tantum enim in plantis naturae germanitas potest, ut non solum saepe praedictos praestet effectus, sed etiam ut alteram in alteram stirpem facile vertat, ut cassiam in cinnamomum, sisymbrium in mentham, triticum in lolium, hordeum in avenam, et ocymum in serpyllum."

I do not know who first emphasized the need for a clear understanding of the sense in which the term species is to be applied. In the second half of the seventeenth century Ray shows some degree of concern on this matter. In the introduction to the *Historia Plantarum*, 1686, he discusses some of the difficulties and lays down the principle that varieties which can be produced from the seed of the same plant are to be regarded as belonging to one species, being, I believe, the first to suggest this definition. That new species can come into existence he denies as inconsistent with Genesis 2, in which it is declared that God finished the work of Creation in six days. Nevertheless he does not wholly discredit the possibility of a "transmutation" of species, such that one species may as an exceptional occurrence give rise by seed to another and nearly

[2] Matthioli Opera, Ed. 1598, p. 8, originally published 1565.

allied species. Of such a phenomenon he gives illustrations the authenticity of which he says he is, against his will, compelled to admit. He adds that some might doubt whether in the cases quoted the two forms concerned are really distinct species, but the passage is none the less of value for it shews that the conception of species as being distinct unchangeable entities was not to Ray the dogma sacrosanct and unquestionable which it afterwards became.[3]

In the beginning of the eighteenth century Marchant,[4] having observed the sudden appearance of a lacinated variety of *Mercurialis*, makes the suggestion that species in general may have arisen by similar mutations. Indeed from various passages it is manifest that to the authors of the seventeenth and early eighteenth centuries species appeared simply as groups more or less definite, the boundaries of which it was unnecessary to determine with great exactitude. Such views were in accord with the general scientific conception of the time. The mutability of

[3] Ray's instances relate to Kales, and in most of these examples we can see that there was no question of mutation or transmutation at all, but that the occurrence was due either to mistake or to cross-fertilisation. Sharrock, to whom Ray refers, was inclined to discredit stories of transmutation, but he has also this passage (*History of the Propagation and Improvement of Vegetables by the Concurrence of Art and Nature*, Oxford, 1660, p. 29):

"It is indeed growen to be a great question, whether the transmutation of a species be possible either in the vegetable, Animal, or Minerall Kingdome. For the possibility of it in the vegetable; I have heard *Mr. Bobart* and his *Son* often report it, and proffer to make oath that the Crocus and Gladiolus, as likewise the Leucoium, and Hyacinths by a long standing without replanting have in his garden changed from one kind to the other: and for satisfaction about the curiosity in the presence of *Mr. Boyle* I took up some bulbs of the very numericall roots whereof the relation was made, though the alteration was perfected before, where we saw the diverse bulbs growing as it were on the same stoole, close together, but no bulb half of the one kind, and the other half of the other: But the changetime being past it was reason we should believe the report of good artists in matters of their own faculty."

Robert Sharrock was a fellow of New College, Oxford. Both the Bobarts were professional botanists, the father was author of a Catalogue of the plants in the Hortus Medicus at Oxford, and the son was afterwards Curator of the Oxford Garden.

[4] *Mém. Ac. roy. des Sci.* for 1719 (1721), p. 59.

species is for example sometimes likened (see for instance Sharrock, loc. cit.) to the metamorphoses of insects, and it is to be
remembered that the search for the Philosopher's Stone by which
the transmutation of metals was to be effected had only recently
fallen into discredit as a pursuit.

The notion indeed of a peculiar, fixed meaning to be attached
to species as distinct from variety is I think but rarely to be
found categorically expressed in prae-Linnaean writings.

But with the appearance of the *Systema Naturae* a great
change supervened. Linnaeus was before all a man of order.
Foreseeing the immense practical gain to science that must come
from a codification of nomenclature, he invented such a system.

It is not in question that Linnaeus did great things for us and
made Natural History a manageable and accessible collection of
facts instead of a disorderly heap; but orderliness of mind has
another side, and inventors and interpreters of systems soon attribute to them a force and a precision which in fact they have not.

The systematist is primarily a giver of names, as Ray with
his broader views perceived. Linnaeus too in the exordium to
the *Systema Naturae* naively remarks, that he is setting out to
continue the work which Adam began in the Golden Age, to give
names to the living creatures. Naming however involves very
delicate processes of mind and of logic. Carried out by the light
of meagre and imperfect knowledge it entails all the mischievous
consequences of premature definition, and promotes facile
illusions of finality. So was it with the Linnaean system. An
interesting piece of biological history might be written respecting
the growth and gradual hardening of the conception of Species.
To readers of Linnaeus's own writings it is well known that his
views cannot be summarized in a few words. Expressed as they
were at various times during a long life and in various connexions,
they present those divers inconsistencies which commonly
reflect a mind retaining the power of development. Nothing
certainly could be clearer than the often quoted declaration of the
Philosophia Botanica, "Species tot numeramus quot diversae
formae in principio sunt creatae," with the associated passage
"Varietates sunt plantae ejusdem speciei mutatae a caussa

quacunque occasionali." Those sayings however do not stand alone. In several places, notably in the famous dissertation on the peloric *Linaria* he explicitly contemplates the possibility that new species may arise by crossing, declaring nevertheless that he thinks such an event to be improbable. In that essay he refers to Marchant's observation on a laciniate *Mercurialis*, but though he states clearly that that plant should only be regarded as a variety of the normal, he does not express any opinion that the contemporary genesis of new species must be an impossibility. In the later dissertation on Hybrid Plants he returns to the same topic. Again though he states the belief that species cannot be generated by cross-breedings, he treats the subject not as heretical absurdity but as one deserving respectful consideration.

The significance of the aphorisms that precede the lectures on the Natural Orders is not easy to apprehend. These are expressed with the utmost formality, and we cannot doubt that in them we have Linnaeus's own words, though for the record we are dependent on the transcripts of his pupils.

The text of the first five is as follows:

1. Creator T. O. in primordio vestiit Vegetabile *Medullare* principiis constitutivis diversi *Corticalis* unde tot difformia individua, quot *Ordines* Naturales prognata.

2. *Classicas* has (1) plantas Omnipotens miscuit inter se, unde tot *Genera* ordinum, quot inde plantae.

3. *Genericas* has (2) miscuit Natura, unde tot *Species* congeneres quot hodie existunt.

4. *Species* has miscuit Casus, unde totidem quot passim occurrunt, *Varietates*.

5. Suadent haec (1–4) Creatoris leges a simplicibus ad Composita.

Naturae leges generationis in hybridis.

Hominis leges ex observatis a posteriori.

I am not clear as to the parts assigned in the first sentence respectively to the "*Medulla*" and the "*Cortex*," beyond that Linnaeus conceived that multiformity was first brought about by diversity in the "*Cortex*." The passage is rendered still more obscure if read in connection with the essay on "*Generatio Ambigena*," where be expresses the conviction that the *Medulla* is contributed by the mother, and the *Cortex* by the father, both in plants and animals.[5]

But however that may be, he regards this original diversity as resulting in the constitution of the Natural Orders, each represented by one individual.

In the second aphorism the Omnipotent is represented as creating the genera by intermixing the individual *plantae classicae*, or prototypes of the Natural Orders.

The third statement is the most remarkable, for in it he declares that Species were formed by the act of Nature, who by inter-mixing the genera produced *Species congeneres*, namely species inside each genus, to the number which now exist. Lastly, Chance or Accident, intermixing the species, produced as many varieties as there are about us.

Linnaeus thus evidently regarded the intermixing of an originally limited number of types as the sufficient cause of all subsequent diversity, and it is clear that he draws an antithesis between *Creator*, *Natura*, and *Casus*, assigning to each a special part in the operations. The acts resulting in the formation of genera are obviously regarded as completed within the days of the Creation, but the words do not definitely show that the parts played by Nature and Chance were so limited.

Recently also E. L. Greene[6] has called attention to some curious utterances buried in the *Species Plantarum*, in which Linnaeus refers to intermediate and transitional species, using language that even suggests evolutionary proclivities of a modern kind, and it is not easy to interpret them otherwise.

Whatever Linnaeus himself believed to be the truth, the effect of his writings was to induce a conviction that the species

[5] *Amoen. Acad.*, 1789, vol. 6. I do not know whether attention has been called to the curious mistake which Linnaeus makes in the course of this argument. He cites the differences between the Mule and the Hinny in illustration of his thesis, pointing out that the Mule is externally more like a horse and the Hinny more like an ass. This, he says, is because the Mule has the horse for a father, and the Hinny the ass, thus inverting the actual facts!

[6] *Proc. Washington Ac. Sci.*, 1909, XI, pp. 17-26.

of animals and plants were immutably fixed. Linnaeus had reduced the whole mass of names to order and the old fantastical transformations with the growth of knowledge had lapsed into discredit; the fixity of species was taken for granted, but not till the overt proclamation of evolutionary doctrine by Lamarck do we find the strenuous and passionate assertions of immutability characteristic of the first half of the nineteenth century.

It is not to be supposed that the champions of fixity were unacquainted with varietal differences and with the problem thus created, but in their view these difficulties were apparent merely, and by sufficiently careful observation they supposed that the critical and permanent distinctions of the true species could be discovered, and the impermanent variations detected and set aside.

This at all events was the opinion formed by the great body of naturalists at the end of the eighteenth and beginning of the nineteenth centuries, and to all intents and purposes in spite of the growth of evolutionary ideas, it remains the guiding principle of systematists to the present day. There are 'good species' and 'bad species' and the systematists of Europe and America spend most of their time in making and debating them.

In some of its aspects the problem of course confronted earlier naturalists. Parkinson for instance (1640) in introducing his treatment of *Hieracium* wrote, "To set forth the whole family of the Hawkeweedes in due forme and order is such a world of worke that I am in much doubt of mine own abilitie, it having lyen heavie on his shoudiers that hath already waded through them . . . for such a multitude of varieties in forme pertaining to one herbe is not to be found againe in *rerum natura* as I thinke," and the same idea, that the difficulty lay rather in man's imperfect powers of discrimination than in the nature of the materials to be discriminated, is reflected in many treatises early and late.

It was however with the great ouburst of scientific activity which followed Linnaeus that the difficulty became acute. Simultaneously vast masses of new material were being collected from all parts of the world into the museums, and the products of the older countries were re-examined with a fresh zeal and on a scale of quantity previously unattempted. But the problem how to name the forms and where to draw lines, how much should be included under one name and where a new name was required, all this was felt, rather as a cataloguer's difficulty than as a physiological problem. And so we still hear on the one hand of the confusion caused by excessive "splitting" and subdivisions, and on the other of the uncritical "lumpers" who associate together under one name forms which another collector or observer would like to see distinguished.

In spite of Darwin's hopes, the acceptance of his views has led to no real improvement—scarcely indeed to any change at all in either the practice or aims of systematists. In a famous passage in the *Origin* he confidently declares that when his interpretation is generally adopted "Systematists will be able to pursue their labours as at present; but they will not be incessantly haunted by the shadowy doubt whether this or that form be a true species. This, I feel sure, and I speak after experience, will be no slight relief. The endless disputes whether or not some fifty species of British brambles are good species will cease." Those disputes nevertheless proceed almost exactly as before. It is true that biologists in general do not, as formerly, participate in these discussions because they have abandoned systematics altogether; but those who are engaged in the actual work of naming and cataloguing animals and plants usually debate the old questions in the old way. There is still the same divergence of opinion and of practice, some inclining to make much of small differences, others to neglect them.

Not only does the work of the systematists as a whole proceed as if Darwin had never written but their attitude towards these problems is but little changed. In support of this statement I may refer to several British Museum Catalogues, much of the *Biologia Centrali-Americana*, Ridgway's *Birds of North America*, the *Fauna Hawaiensis*, indeed to almost any of the most important systematic publications of England, America, or any other country. These works are compiled by the most proficient

systematists of all countries in the several groups, but with
rare exceptions they show little misgiving as to the fundamental
reality of specific differences. That the systematists consider
the species-unit as of primary importance is shown by the
fact that the whole business of collection and distribution of
specimens is arranged with regard to it.

Almost always the collections are arranged in such a way that
the phenomena of variation are masked. Forms intermediate
between two species are, if possible, sorted into separate boxes
under a third specific name.   If a species is liable to be constantly
associated with a mutational form, the mutants are picked out,
regardless of the circumstances of their origin, from the samples
among which they were captured, and put apart under a special
name.   Only by a minute study of the original labels of the
specimens and by redistributing them according to locality and
dates, can their natural relations be traced. The published
accounts of these collections often take no notice of variations,
others make them the subject of casual reference. Very few
indeed treat them as of much importance. From such indications it is surely evident that the systematists attach to the
conception of species a significance altogether different from that
which Darwin contemplated.

I am well aware that some very eminent systematists regard
the whole problem as solved. They hold as Darwin did that
specific diversity has no physiological foundation or causation
apart from fitness, and that species are impermanent groups,
the delimitations of which are ultimately determined by environmental exigency or "fitness." The specific diversity of
living things is thus regarded as being something quite different
in nature from the specific diversity of inorganic substances.
In practice those who share these opinions are, as might be anticipated, to be found among the 'lumpers' rather than among
the 'splitters.' In their work, certainly, the Darwinian theory
is actually followed as a guiding principle; unanalysed inter-
gradations of all kinds are accepted as impugning the integrity
of species; the underlying physiological problem is forgotten,
and while the product is amost valueless as a contribution to

biological research, I can scarcely suppose that it aids greatly in the advances of other branches of our science.

But why is it that, with these exceptions, the consequences of the admittedly general acceptance of a theory of evolution are so little reflected in the systematic treatment of living things? Surely the reason is that though the systematist may be convinced of the general truth of the evolution theory at large, he is still of opinion that species are really distinct things. For him there are still 'good' species and 'bad' species and his experience tells him that the distinction between the two is not simply a question of degree or a matter of opinion.

To some it may seem that this is mere perversity, a refusal to see obvious truth, a manifestation of the spirit of the collector rather than of the naturalist. But while recognising that from a magnification of the conception of species the systematists are occasionally led into absurdity I do not think the grounds for their belief have in recent times been examined with the consideration they deserve. The phenomenon of specific diversity is manifested to a similar degree by living things belonging to all the great groups, from the highest to the lowest, Vertebrates, Invertebrates, Protozoa, Vascular Plants, Algae, and Bacteria, all present diversities of such a kind that among them the existence of specific differences can on the whole be recognised with a similar degree of success and with very similar limitations. In all these groups there are many species quite definite and unmistakable, and others practically indefinite. The universal presence of specificity, as we may call it, similarly limited and characterised, is one of its most remarkable features. Not only is this specificity thus universally present among the different forms of life, but it manifests itself in respect of the most diverse characteristics which living things display. Species may thus be distinguished by peculiarities of form, of number, of geometrical arrangement, of chemical constitution and properties, of sexual differentiation, of development, and of many other properties. In any one or in several of these features together, species may be found distinguished from other species. It is also to be observed that the definiteness of these distinctions has no essential dependence on the nature of the characteristic which manifests them. It is for example sometimes said that colour-distinctions are of small systematic importance, but every systematist is familiar with examples (like that of the wild species of *Gallus*) in which colours though complex, show very little variation. On the other hand features of structure, sexual differentiation, and other attributes which by our standards are estimated as essential, may be declared to show much variation or little, not according to any principle which can be detected, but simply as the attention happens to be applied to one species or group of species, or to another. In many groups of animals and plants observers have hit upon characters which were for a time thought to be finally diagnostic of species. The Lepidoptera and Diptera for instance, have been re-classified according to their neuration. Through a considerable range of forms determinations may be easily made on these characters, but as is now well known, neuration is no more immune from variation than any other feature of organisation, and in some species great variability is the rule. Again it was once believed by some that the genitalia of the Lepidoptera provided a basis of final determination — with a similar sequel. In some groups, for example the Lycaenidae, or the Hesperidae, there are forms almost or quite indistinguishable on external examination, but a glance at the genitalia suffices to distinguish numerous species, while on the contrary among Pieridae a great range of species show scarcely any difference in these respects: and again in occasional species the genitalia show very considerable variations.

The proposition that animals and plants are on the whole divisible into definite and recognisable species is an approximation to the truth. Such a statement is readily defensible, whereas to assert the contrary would be palpably absurd. For example, a very competent authority lately wrote: "In the whole Lepidopterous fauna of England there is no species of really uncertain limits." [7] Others may be disposed to make certain reservations, but such exceptions would be so few as scarcely to impair the validity of the general statement. The

[7] J. W. Tutt, in *Ent. Rec.*, 1909, XXI, p. 185.

declaration might be extended to other orders and other lands.

We know, of course, that the phenomenon of specific diversity is complicated by local differentiation: that, in general, forms which cannot disperse themselves freely exhibit a multitude of local races, and that of these some are obviously adaptative, and that a few even owe their peculiarity to direct envitonmental effects. Every systematist also is perfectly aware that in dealing with collections from little explored countries the occurrence of polymorphism or even of sporadic variation may make the practical business of distinguishing the species difficult and perhaps for the time impossible; still, conceding that a great part of the diversity is due to geographical differentiation, and that some is sporadic variation, our experience of our own floras and faunas encourages the belief that if we were thoroughly familiar with these exotic productions it would usually be possible to assign their specific limitations with an approach to certainty.

For apart from any question of the justice of these wider inferences, if we examine the phenomenon of specificity as it appears in those examples which are nearest to hand, surely we find signs in plenty that specific distinction is no mere consequence of Natural Selection. The strength of this proposition has lain mainly in the appeal to ignorance. Steadily with the growth of knowledge has its cogency diminished, and such a belief could only have been formulated at a time when the facts of variation were unknown.

In Darwin's time no serious attempt had been made to examine the manifestations of variability. A vast assemblage of miscellaneous facts could formerly be adduced as seemingly comparable illustrations of the phenomenon "Variation." Time has shown this mass of evidence to be capable of analysis. When first promulgated it produced the impression that variability was a phenomenon generally distributed amongst living things in such a way that the specific divisions must be arbitrary. When this variability is sorted out, and is seen to be in part a result of hybridisation, in part a consequence of the persistence of hybrids by parthenogenetic reproduction, a polymorphism due to the continued presence of individuals representing various combinations of Mendelian allelomorphs, partly also the transient effect of alteration in external circumstances, we see how cautious we must be in drawing inferences as to the indefiniteness of specific limits from a bare knowledge that intermediates exist. Conversely, from the accident of collocation or from a misleading resemblance in features we deem essential, forms genetically distinct are often confounded together, and thus the divergence of such forms in their other features, which we declare to be non-essential, passes as an example of variation. Lastly, and this is perhaps the most fertile of all the sources of confusion, the impression of the indefiniteness of species is created by the existence of numerous local forms, isolated geographically from each other, forms whose differences may be referable to any one of the categories I have enumerated.

The advance has been from many sides. Something has come from the work of systematists, something from cultural experiments, something from the direct study of variation as it appears in nature, but progress is especially due to experimental investigation of heredity. From all these lines of inquiry we get the same answer; that what the naturalists of fifty years ago regarded as variation is not one phenomenon but many, and that what they would have adduced as evidence against the definiteness of species may not in fact be capable of this construction at all.

If we may once more introduce a physical analogy, the distinctions with which the systematic naturalist is concerned in the study of living things are as multifarious as those by which chemists were confronted in the early days of their science. Diversities due to mechanical mixtures, to allotropy, to differences of temperature and pressure, or to degree of hydration, had all to be severally distinguished before the essential diversity due to variety of chemical constitution stood out clearly, and I surmise that not till a stricter analysis of the diversities of animals and plants has been made on a comprehensive scale, shall we be in a position to declare with any confidence whether there is or is not a natural and physiological distinction between species and variety.

As I have said above, it is in the cases nearest to hand that the problem may be most effectively studied. Comparison between forms from dissimilar situations contributes something; but it is by a close examination of the behaviour, especially the genetic behaviour, of familiar species when living in the presence of their nearest allies that the most direct light on the problem is to be obtained. I cannot understand the attitude of those who, contemplating such facts as this examination elicits, can complacently declare that specific difference is a mere question of degree. With the spread of evolutionary ideas to speak much of the fixity of species has become unfashionable, and yet how striking and inscrutable are the manifestations of that fixity!

Consider the group of species composing the *agrestis* section of the genus *Veronica*, namely *Tournefortii*, *agrestis*, and *polita*.

These three grow side by side in my garden, as they do in suitable situations over a vast area of the temperate regions. I have for years noticed them with some care and become familiar with their distinctions and resemblances. Never is there any real doubt as to the identity of any plant. The species show some variability, but I have never seen one which assumed any of the distinguishing features of the others. A glance at the fruits decides at once to which species a plant belongs. I find it impossible to believe that the fixity of these distinctions is directly dependent on their value as aids in the struggle for existence. The mode of existence of the three forms in so far as we can tell is closely similar. By whatever standard we reckon systematic affinity I suppose we shall agree that these species come very near indeed to each other. Bentham even takes the view that *polita* is a mere variety of *agrestis*.

Now in such cases as this it has been argued that the specific features of the several types have been separately developed in as many distinct localities, and that their present association is due to subsequent redistribution. Of these Veronicas indeed we know that one, *Tournefortii* (=*Buxbaumii*) is as a matter of fact a recent introduction from the east.[8] But this course of argument leads to still further difficulties. For if it is true that the peculiarities of the several species have been perfected and preserved on account of their survival-value to their possessors, it follows that there must be many ways of attaining the same result. But since sufficient adaptation may be ensured in so many ways, the disappearance of the common parent of these forms is difficult to understand. Obviously it must have been a plant very similar in general construction to its modern representatives. Like them it must have been an annual weed, with an organisation conformable to that mode of life. Why then, after having been duly perfected for that existence should it have been entirely superseded in favour of a number of other distinct contrivances for doing the same thing, and—if a gradual transition be predicated—not only by them, but by each intermediate stage between them and the original progenitor? Surely the obvious inference from such facts is that the burden cast upon the theory of gradual selection is far greater than it can bear; that adaptation is not in practice a very close fit, and that the distinctions between these several species of Veronica have not arisen on account of their survival-value but rather because none of their diversities was so damaging as to lead to the extermination of its possessor. When we see these various Veronicas each rigidly reproducing its parental type, all comfortably surviving in competition with each other, are we not forced to the conclusion that *tolerance* has as much to do with the diversity of species as the stringency of Selection? Certainly these species owe their continued existence to the fact that they are each good enough to live, but how shall we refer the distinctions between them directly or indirectly to the determination of Natural Selection?

[8] E. Lehmann (*Bull. l'Herb. Boissier*, Ser. 2, VIII, 1908, p. 229) has published an admirable paper on the interrelationships of these species and has instituted cultural experiments which will probably much elucidate the nature of their specific distinctness. As regards the existence of intermediate forms he comes to the conclusion that two only can be so regarded. The first was described by Kuntze from specimens found on a flower-pot on board a Caspian steamer, from which Lehmann proposes the new specific name *Siaretensis*. This comes between *polita* and *filiformis*, a close ally of *Tournefortii*. The other, which combines some of the features of both *polita* and *Tournefortii*, was found in the province of Asterabad.

3

The control of Selection is loose while the conformity to specific distinction is often very strict and precise, and no less so even when several closely related species co-exist in the same area and in the same circumstances.

The theory of Selection fails at exactly the point where it was devised to help: *Specific* distinction.

Let us examine a somewhat different set of facts in the case of another pair of nearly allied species *Lychnis diurna* and *vespertina*. The two plants have much in common. Both are dioecious perennials, with somewhat similar flowers, the one crimson, the other white. Each however has its peculiarities which are discernible in almost any part of its structure, whether flower, leaf, fruit or seed, distinctions which would enable a person thoroughly familiar with the plants to determine at once from which species even a small piece had been taken. There is so much resemblance however as readily to support the surmise that the two were mere varieties of one species. Bentham, following Linnaeus, in fact actually makes this suggestion, with what propriety we will afterwards consider. Now this case is typical of many. The two forms have a wide distribution, occurring sometimes separately, sometimes in juxtaposition. *L. diurna* is a plant of hedgerows and sheltered situations. *L. vespertina* is common in fields and open spaces, where *diurna* is hardly ever found; but not rarely *vespertina* occurs in association with *diurna* in the places which that plant frequents. In this case I do not doubt that we have to do with organisms of somewhat different aptitudes. That *L. vespertina* has powers which *diurna* has not is shown very clearly by the fact that *diurna* is sometimes entirely absent from areas where *vespertina* can abound.[9] But in order to understand the true genetic relations of the two plants to each other it is necessary to observe their behaviour when they meet as they not unfrequently do.

[9] In Cambridgeshire for example *vespertina* is common but *diurna* is absent. Whether this absence is connected with the general presence of chalk I cannot say. When introduced artificially *diurna* establishes itself, for a time at least, without any apparent difficulty and occasionally escapes from the garden on to the neighbouring roadside.

If the *Lychnis* population of such a locality be examined it will
be found to consist of many undoubted and unmodified *diurna*,
a number—sometimes few, sometimes many—of similarly
unmodified *vespertina*, and an uncertain but usually rather small
proportion of plants obviously hybrids between the two. How
is it possible to reconcile these facts with the view that specific
distinction has no natural basis apart from environmental
exigency?

Darwinian orthodoxy suggests that by a gradual process of
Natural Selection either one of these two types was evolved from
the other, or both from a third type. I cannot imagine that
anyone familiar with the facts would propose the first hypothesis
in the case of *Lychnis*, nor can I conceive of any process, whether
gradual or sudden, by which *diurna* could have come out of
vespertina, or *vespertina* out of *diurna*. Both however may no
doubt have been derived from some original third type. It is
conceivable that *Lychnis macrocarpa* of Boissier, a native of
Southern Spain and Morocco, may be this original form. This
species is said to combine a white flower (like that of *L. vespertina*), with capsule-teeth rolled back (like those of *diurna*).[10]
But whatever the common progenitor may have been, if we are
to believe that these two species have been evolved from it by
a gradual process of Natural Selection based on adaptation,
enormous assumptions must be made regarding the special fitness
of these two forms and the special unfitness of the common
parent, and these assumptions must be specially invoked and
repeated for each several feature of structure or habits distinguishing the three forms.

Why, if the common parent was strong enough to live to give
rise to these two species, is it either altogether lost now, or at
least absent from the whole of Northern Europe? Its two
putative descendants, though so distinct from each other, are,
as we have seen, able often to occupy the same ground. If
they were gradually derived from a common progenitor—
necessarily very like themselves—can we believe that this original

[10] Conceivably however it may be a segregated combination. For an account
of this plant see Boissier, *Voy. Bot. Midi de l'Espagne*, 1839, II, 722.

form should always, in all the diversities of soil and situation which they inhabit, be unable to exist? Some one may fancy that the hybrids which are found in the situations occupied by both forms are this original parental species. But nothing can be more certain than that these plants are simply heterozygous combinations made by the union of gametes bearing the characters of *diurna* and *vespertina*.[11] For they may be reproduced exactly in F_1 or in later generations of that cross when it is artificially made; when bred from their families exhibit palpable phenomena of segregation more or less complex; and usually, if perhaps not always, they are partially sterile.[12] In a locality on the Norfolk coast that I know well, there is a strip of rough ground chiefly sand-bank, which runs along the shore. This ground is full of *vespertina*. Not a hundred yards inland is a lane containing *diurna*, and among the *vespertina* on the sand-bank are always some of the hybrid form, doubtless the result of fertilisation from the heighbouring *diurna* population. Seed saved from these hybrids gave *vespertina* and hybrids again, having obviously been fertilised by other *vespertina* or by other hybrids, and I have no doubt that such hybrid plants if fertilised by *diurna* would have shown some *diurna* offspring. The absence of *diurna* in such localities may fairly be construed as an indication that *diurna* is there at a real disadvantage in the competition for life.

But if, admitting this, we proceed to consider how the special aptitude of *vespertina* is constituted, or what it is that puts *diurna* at a disadvantage, we find ourselves quite unable to show the slightest connexion between the success of one or the

[11] A discussion of this subject with references to literature is given by Rolfe, in an excellent paper on "Hybridisation viewed from the standpoint of Systematic Botany" (*Jour. R. Hort. Soc.*, XXIV, 1900, p. 197). He concludes: "The simple fact is that the two plants (*L. diurna* and *vespertina*) are thoroughly distinct in numerous particulars, and affect such different habitats that in some localities one or the other of them is completely wanting. But when their stations are adjacent they hybridise together very readily, and it is here that these intermediate forms occur which have puzzled botanists so much." The same paper contains valuable information concerning several cognate illustrations.

[12] In only two cases have I seen such plants (both females) completely sterile.

failure of the other on the one hand, and *the specific character-istics* which distinguish the two forms on the other. The orthodox Selectionist would, as usual, appeal to ignorance. We ask what can *vespertina* gain by its white flowers, its more lanceolate leaves, its grey seeds, its almost erect capsule-teeth, its longer fruits, which *diurna* loses by reason of its red flowers, more ovate leaves, dark seeds, capsule-teeth rolled back, and shorter fruits? We are told that each of these things *may* affect the viability of their possessors. We cannot assert that this is untrue, but we should like to have evidence that it is true. The same problem confronts us in thousands upon thousands of examples, and as time goes on we begin to feel that speculative appeals to ignorance, though dialectically admissible, provide an insufficient basis for a proposition which, if granted, is to become the foundation of a vast scheme of positive construction.

One thing must be abundantly clear to all, that to treat two forms so profoundly different as one, because intermediates of unknown nature can be shown to exist between them, is a mere shirking of the difficulties, and this course indeed creates artificial obstacles in the way of those who are seeking to discover the origin of organic diversity.

In the enthusiasm with which evolutionary ideas were received the specificity of living things was almost forgotten. The exactitude with which the members of a species so often conform in the diagnostic, specific features passed out of account; and the scientific world by dwelling with a constant emphasis on the fact of variability, persuaded itself readily that species had after all been a mere figment of the human mind. Without presuming to declare what future research only can reveal, I anticipate that, when variation has been properly examined and the several kinds of variability have been successfully distinguished according to their respective natures, the result will render the natural definiteness of species increasinlgy apparent. Formerly in such a case as that of the two *Lychnis* species, the series of "intermediates" was taken to be a palpable proof that *vespertina* "graded" to *diurna*. It is this fact, doubtless, upon which Bentham would have relied in suggesting that both may be one species.[13] Genetic tests, though as yet imperfectly applied, make it almost certain that these intergrading forms are not in any true sense variations from either species in the direction of the other, but combinations of elements derived from both.

The points in which very closely allied species are distinguished from each other may be found in the most diverse features of their organisation. Sometimes specific difference is to be seen in a character which we can believe to be important in the struggle, but at least as often it is some little detail that we cannot but regard as trivial which suffices to differentiate the two species. Even when the diagnostic point is of such a nature that we can imagine it to make a serious difference in the economy we are absolutely at a loss to suggest why this feature should be a necessity to species A and unnecessary to species B its nearest ally. The house sparrow (*Passer domesticus*) is in general structure very like the tree sparrow (*P. montanus*). They differ in small points of colour. For instance *montanus* has a black patch on the cheek which is absent in *domesticus*. The presence in the one species and the absence in the other are equally definite, and in both cases we are equally unable to suggest any consideration of utility in relation to these features. The two species are distinguished also by a characteristic that may well be supposed to be of great significance. In *domesticus* the two sexes are strongly differentiated, the cock being more ornate than the hen. On the other hand the two sexes in *montanus* are alike, and, if we take a standard from *domesticus*, we may fairly say that in *montanus* the hen has the colouration of the male. It is not unreasonable to suppose that such a distinction may betoken some great difference in physiological economy, but the economical significance of this perhaps important distinction is just as unaccountable as that of the seemingly trivial but equally diagnostic colour-point.

I have spoken of the fixed characteristics of the two species.

[13] As is well known, in an even more notorious example, he proposed to unite *Primula vulgaris*, *P. elatior*, and *P. acaulis*, similarly relying on the existence of "intermediates," which we now well know to be mongrels between the species.

If we turn to a very different feature, their respective liability to albinistic variation, we find ourselves in precisely similar difficulty. *Passer domesticus* is a species in which individuals more or less pied occur with especial frequency, but in *P. montanus* such variation is extremely rare if it occurs at all. The writer of the section on Birds in the *Royal Natural History* (III., 1894-5, p. 393) calls attention to this fact and remarks that in that species he knows no such instance.

The two species therefore, apart from any differences that we can suppose to be related to their respective habits, are characterised by small fixed distinctions in colour-markings, by a striking difference in secondary sexual characters, and by a difference in variability. In all these respects we can form no surmise as to any economic reason why the one species should be differentiated in the one way and the other in the other way, and I believe it is mere self-deception which suggests the hope that with fuller knowledge reasons of this nature would be discovered.

The two common British wasps, *Vespa vulgaris* and *Vespa germanica*, are another pair of species closely allied although sharply distinguished, which suggest similar reflexions. Both usually make subterranean nests but of somewhat different materials. *V. vulgaris* uses rotten wood from which the nest derives a characteristic yellow colour, while *V. germanica* scrapes off the weathered surfaces of palings and other exposed timber, material which is converted into the grey walls of the nest. The stalk by which the nest is suspended (usually to a root) in the case of *germanica* passes freely through a hole in the external envelope, but *vulgaris* unites this external wall solidly to the stalk. In bodily appearance and structure the two species are so much alike that they have often been confounded even by naturalists, and to the untrained observer they are quite indistinguishable. There are nevertheless small points of difference which almost though not quite always suffice to distinguish the two forms. For example the yellow part of the sinus of the eyes is emarginate in *vulgaris* but not emarginate in *germanica*. *V. vulgaris* often has black spots on the tibiae while in *germanica* the

tibiae are usually plain yellow. In both species there is a hori-
zontal yellow stripe on the thorax, but whereas in *vulgaris* this is
a plain narrow stripe, it is in *germanica* enlarged downwards in
the middle. These and other apparently trivial details of colouration, though not absolutely constant, are yet so nearly constant
that irregularities in these respects are quite exceptional. Lastly
the genitalia of the males, though not very different, present
small structural points of distinction which are enough to distinguish the two species at a glance.[14]

In considering the meaning of the distinctions between these
two wasps we meet the old problem illustrated by the Sparrows.
The two species have somewhat different habits of life and we
should readily expect to find differences of bodily organisation
corresponding with the differences of habits. But is that what
we do find? Surely not. To suppose that there is a correspondence between the little points of colour and structure which
we see and the respective modes of life of the two species is
perfectly gratuitous. We have no inkling of the nature of such
a correspondence, how it can be constituted, or in what it may
consist.

Is it not time to abandon these fanciful expectations which
are never realised? Everywhere both among animals and plants
does the problem of specific difference reiterate itself in the same
form. In view of such facts as I have related and might indefinitely multiply, the fixity of specific characters cannot readily
be held to be a measure of their economic importance to their
possessors. The incidence of specific fixity is arbitrary and
capricious, sometimes lighting on a feature or a property which
can be supposed to matter much, but as often is it attached to the
most trifling of superficial peculiarities.

The incidence of *variability* is no less paradoxical, and without
investigation of the particular case no one can say what will be

[14] For an account of the distinctions between *Vespa vulgaris* and *germanica*
see Ch. Janet, *Etudes sur les Fourmis, les Guêpes et les Abeilles*, 11°, Note. Sur
Vespa germanica et *V. vulgaris*. Limoges (Ducourtieux), 1895; and R. du Buysson,
Monographie des Guepes, *Ann. Soc. Ent. France*, 1903, Vol. LXXII, p. 603, Pl.
VIII.

found to show much or little variability. The very characteristic which in one species may exhibit extreme variability
may in an allied species show extreme constancy. Illustrations
will occur to any naturalist, but nowhere is this truth more
strikingly presented than in the British Noctuid Moths. Many
are so variable that, in the common phrase, "scarcely two can
be found alike," while others show comparatively slight variation.
It need scarcely be remarked that, in the instances I have in
mind, the evidence of great variability is in no way due to the
abundance with which the particular species occurs, for common
species may show constancy, and less abundant species may show
great variability. The polymorphism seems to be now at least
a general property of the variable species, as the fixity is a
property of the fixed species. In illustration I may refer to the
following examples.

Dianthoecia capsincola is a common and widely distributed
moth which feeds on *Lychnis*. It shows little variation. *Dianthoecia carpophaga* is another species which feeds chiefly on
Silene. Its habits are very similar to those of *capsincola*. Like
that species it has a wide geographical range and is abundant
in its localities, but in contrast to the fixity of *capsincola*, *carpophaga* exhibits a complex series of varieties. *Agrotis suffusa*
(= *ypsilon*) is a moth widely spread through the southern half
of England. It is very constant in colour and markings. *Agrotis segetum* and *tritici* are excessively variable both in ground
colour and markings, being found in an immense profusion of
dissimilar forms throughout their distribution. Of these and
several other species of *Agrotis* there are many named varieties,
some of which have by various writers been regarded as speci-
fically distinct. Of the genus *Noctua* many species (e. g. *festiva*)
show a similar polymorphism, but *N. triangulum*, though showing
some variation in certain respects, is usually very constant to
its type, and the same is true of *N. umbrosa*.

In several species of *Taeniocampa*, especially *instabilis*, the
multiplicity of forms is extreme, while *cruda* (= *pulverulenta*)
is a comparatively constant species. The genus *Plusia* contains
a number of constant species, but in *Plusia interrogationis* we

meet the fact that the central silvery mark undergoes endless variation. "Truly no two are alike," says Mr. Tutt, "and to look down a long series of *interrogationis* is something like looking at a series of Chinese characters." In contrast to this we have the fact that in *Plusia gamma* the very similar silvery mark is by no means variable.

I have taken this series of cases from the Noctuid moths, but it would be as easy to illustrate the same proposition from the Geometridae or the Micro-Lepidoptera.[15] I have a long series of *Peronea cristana*, for example, which was given to me by Mr. W. H. B. Fletcher, of Bognor. All were beaten out of the same hedge, and their polymorphism is such that no one unaccustomed to such examples could suppose that they belonged to a single species. Another common form, *P. schalleriana*, which lives in similar circumstances, exhibits comparatively slight variability.

It should be expressly noted that the variation of which I am speaking is a genuine polymorphism. Several of the species enumerated exhibit also geographical variation, possessing definite and often strikingly distinct races peculiar to certain localities; but apart from the existence of such local differentiation, stands out the fact upon which I would lay stress, that some species are excessively variable while others are by comparison constant, in circumstances that we may fairly regard as comparable.

This fact is difficult to reconcile with the conventional view that specific type is directly determined by Natural Selection

[15] The statements made above are for the most part taken from Barrett, C. G., *Lepidoptera of the British Islands*, and from Tutt, J. W., *The British Noctuae and their Varieties*. The reader who is unfamilar with the amazing polymorphism exhibited by some of these moths should if possible take an opportunity of looking over a long series in a collection, or, if that be impossible, refer to the admirable coloured plates published by Barrett. It may not be superfluous to observe that plenty of similar examples are known in other countries. For instance *Plotheia frontalis*, a Noctuid which often abounds in Ceylon, shows an equally bewildering wealth of forms. If a dozen specimens of such a species were to be brought home from some little known country, each individual would almost certainly be described as the type of a distinct species. (See the coloured plate published by Sir G. Hampson, Cat. Brit. Mus., Heterocera, Vol. IX.)

and that the precision with which a species conforms to its pattern is an indication of the closeness of that control. Anyone familiar with the characteristics of Moths will agree that the Noctuids, Geometrids and Tortricids are creatures whose existence depends in some degree on the success with which they can escape detection by their enemies in the imaginal state. We are therefore not surprised to find that some species of these orders exhibit definite geographical variation in conformity with the character of the ground, which may reasonably be supposed to aid in their protection. If this were all, there would be nothing to cause surprise. We might even be disposed to allow that variability might contribute to the perpetuation of animals so situated, on the principle that among a variety of surroundings some would probably be in harmony with the objects on which they rest. But we cannot admit the plausibility of an argument which demands on the one hand that the extreme precision with which species A adheres in the minutest details of its colour and pattern to a certain type shall be ascribed to the protective fitness of those details, and on the other hand that the abundant variability of species B shall be ascribed to the same determination. If it is absolutely necessary for A to conform to one type how comes it that B may range through some twenty distinct forms, any two of which differ more from each other than the regular species of many other genera? The only reply I can conceive is a suggestion that there *may* be some circumstance which differentiates the various classes of cases, that the exigencies of the fixed species *may* be different from those of the variable. Those who make such appeals to ignorance do not always perhaps realise whither this course of reasoning may lead. If admissible here the same argument would lead us to suggest that because albino moles have for an indefinite period occurred on a certain land near Bath there may be something in the soil or in the conditions of life near Bath which requires a proportion of albinos in its mole population. Or again, because the butterfly *Thais rumina* in one locality, Digne in the south of France, has a percentage of individuals of the variety *Honoratii* (with certain normally yellow spots on the hind wing coloured bright red) and nowhere else throughout its distribution, that therefore we may suggest that there is some difference in the condition of life at Digne which makes the continuance of *Honoratii* there possible and beneficial.

A polymorphism offering a parallel to that of the variable moths is afforded by the breeding plumage of the Ruff, the male of *Machetes pugnax*. The variety of plumage which these cocks exhibit is such that the statement that no two can be found alike is only a venial exaggeration. Newton remarks[16] "that all this wonderful 'show' is the consequence of the polygamous habit of the Ruff can scarcely be doubtful"; but even if it be conceded that the great external differentiation of the cocks may be a result of sexual selection, the problem of their *polymorphism* remains unsolved, for, as we are well aware, polygamy is not usually associated with polymorphism of the male. The Black Cock (*Tetrao tetrix*), for example, is as polygamous as the Ruff, but in that and countless other cases, both sexes are constant to one type of plumage.

When we thus compare the polymorphism of one species with the fixity of another, and attempt to determine the causes which have led to these extraordinary contrasts, two distinct lines of argument are open to us. We may ascribe the difference either to causes external to the organisms, primarily, that is to say, to a difference in the exigencies of Adaptation under Natural Selection; or on the other hand we may conceive the difference as due to innate distinctions in the chemical and physiological constitutions of the fixed and the variable respectively. There is truth undoubtedly in both conceptions. If the mole were physiologically incapable of producing an albino that variety would not have come into being, and if the albino were totally incapable of getting its living it would not be able to hold its

---

[16] *Dict. of Birds*, p. 800. It would be interesting and profitable to attempt in a long series of Ruffs to determine the Mendelian factors which by their combinations give rise to this complex assemblage of varietal forms. A few such factors both of colour and pattern can be at once distinguished, and it is noticeable that some of the resulting types of barring, spangling and penciling show a perceptible correspondence with some of the types of colouration found in the breeds of domestic fowls.

own. Were *Plotheia frontalis* constructed on a chemical plan which admitted of no variation, the countless varieties would not have been produced; and if one of its varieties had an overwhelming success out of all proportion to that of the rest, then the species would soon become monomorphic again. We cannot declare that Natural Selection has no part in the determination of fixity or variability; nevertheless looking at the whole mass of fact which a study of the incidence of variation provides, I incline to the view that the variability of polymorphic forms should be regarded rather as a thing tolerated than as an element contributing directly to their chances of life; and on the other hand that the fixity of the monomorphic forms should be looked upon not so much as a proof that Natural Selection controls them with a greater stringency, but rather as evidence of a natural and intrinsic stability of chemical constitution. ¹

Compare the condition of a variable form like the male Ruff (or in a less degree the Red Grouse in both its sexes) with that of the common Pheasant which is comparatively constant. In the Pheasant no doubt variations do occur as in other wild birds, but apart from the effects of mongrelisation the species is unquestionably uniform. Could it seriously be proposed that we should regard the constancy of the pheasant's plumage in this country as depending on the special fitness of that type of colouration? Even if the pheasant be not an alien in Western Europe, it has certainly been protected for centuries, and for a considerable period has existed in a state of semi-domestication. Such conditions should give good opportunity for polymorphism to be produced. In some coverts various aberrations do of course occur and persist, yet there is nothing indicative of a general relaxation of the fixity of the specific type, and the pheasant remains substantially a fixed species.¹ The common pheasant (*Phasianus colchicus*) even shows little of that disposition to

¹ Howard Saunders (*Illust. Manual of British Birds*, 1899, p. 499) states that there is evidence that the pheasant had become naturalized in the south of England before the Norman invasion. He adds, "little, if any, deviation from the typical *P. colchicus* took place up to the end of last century, when the introduction of the Chinese Ring-necked *P. torquatus* commenced, which has left almost indelible marks, especially with regard to the characteristic white collar."

form local races which appears in the species of Further India. Are we not then on safer ground in regarding the fixity of our species as a property inherent in its own nature and constitution? Just as in ages of domestication no rose has ever given off a blue variety so has the pheasant never broken out into the polymorphism of the Ruff.

As soon as it is realised how largely the phenomena of variation and stability must be an index of the internal constitution of organisms, and not mere consequences of their relations to the outer world, such phenomena acquire a new and more profound significance.

CHAPTER II

Twenty years ago in describing the facts of Variation, argument was necessary to show that these phenomena had a special value in the sciences of Zoology and Botany. This value is now universally understood and appreciated. In spite however of the general attention devoted to the study of Variation, and the accumulation of material bearing on the problem, no satisfactory or searching classification of the phenomena is possible. The reason for this failure is that a real classification must presuppose knowledge of the chemistry and physics of living things which at present is quite beyond our reach.

It is however becoming probable that if more knowledge of the chemical and physical structure of organisms is to be attained, the clue will be found through Genetics, and thus that even in the uncoordinated accumulation of facts of Variation we are providing the means of analysis applicable not only to them, but to the problems of normality also.

The only classification that we can yet institute with any confidence among th phenomena of Variation is that which distinguishes on the one hand variations in the processes of division from variations in the nature of the substances divided.

Variations in the processes of division are most often made apparent by a change in the number of the parts, and are therefore called *Meristic* Variations, while the changes in actual composition of material are spoken of as *Substantive* Variations. The Meristic Variations form on the whole a natural and fairly well defined group, but the Substantive Variations are obviously a heterogeneous assemblage.

Though this distinction does not go very far, it is useful, and in all probability fundamental. It is of value inasmuch as it brings into prominence the distinct and peculiar part which the process of division, or, more generally, repetition of parts, plays in the constitution of the forms of living things.

That there may be a real independence between the Meristic and the Substantive phenomena is evident from the fact both that Meristic changes may occur without Substantive Variation, and that the substances composing an organism may change without any perceptible alteration in its meristic structure. When the distinction between these two classes of phenomena is perceived it will be realised that the study of genetics has on the one hand a physical, or perhaps more strictly a mechanical aspect, which relates to the manner in which material is divided and distributed; and also a chemical aspect, which relates to the constitution of the materials themselves. Somewhat as the philosophers of the seventeenth and eighteenth centuries were awaiting both a chemical and a mechanical discovery which should serve as a key to the problems of unorganised matter, so have biologists been awaiting two several clues. In Mendelian analysis we have now, it is true, something comparable with the clue of chemistry, but there is still little prospect of penetrating the obscurity which envelops the mechanical aspect of our phenomena. To make clear the application of the terms chemical and mechanical to the problem of Genetics the nature of that problem must be more fully described. In its most concrete form this problem is expressed in the question, how does a cell divide? If the organism is unicellular, and the single cell is the whole body, then the process of heredity is accomplished in the single operation of cell-division. Similarly in animals and plants whose bodies are made up of many cells, the whole process of heredity is accomplished in the cell-divisions by which the germ-cells are formed. When therefore we see a cell dividing, we are witnessing the process by which the form and the properties of the daughter-cells are determined.

Now this process has the two aspects which I have called mechanical and chemical. The term "*Entwicklungsmechanik*" has familiarised us with the application of the word mechanics to these processes, but on reflexion it will be seen that this comprehensive term includes two sorts of events which are sometimes readily distinguishable. There is the event by which the cell *divides*, and the event by which the two halves or their descendants are or may be *differentiated*. It is common knowledge that in some cell-divisions two similar halves, indistinguishable in appearance, properties, and subsequent fate, may be produced, while in other divisions daughter-cells with distinct properties and powers are formed. We cannot imagine but that in the first case, when the resulting cells are idenfical, the division is a mechanical process by which the mother-cell is simply cut in two; while in order that two differentiated halves may be produced, some event must have taken place by which a chemical distinction between the two halves is effected.[1] In any ordinary Mendelian case we have a clear proof that such a chemical difference may be established between germ-cells. The facts of colour-inheritance for instance prove that germ-cells, otherwise identical, may be formed *possessing* the chromogen-factor which is necessary to the formation of colour in the flowers, or *destitute* of that factor. Similarly the germ-cells may possess the ferment which, by its action on the chromogenic substance, produces the colour, or they may be without that ferment. The same line of argument applied to a great range of cases. Nevertheless, though differences in chemical properties are often thus constituted by cell-divisions, and though we are thus able to make a quasi-chemical analysis of the individual by determining and enumerating these properties, yet it is evident that the distribution of these factors is not itself a chemical process. This is proved by the fact that similar divisions may be effected between halves which are exactly alike, and also by the fact that the numbers in which the various types of germ-cells are formed negative any suggestion of valency between them. The recognition of the unit-factors may lead—indeed must lead—to great advances in chemical physiology which without that clue would have been impossible, but in causation the chemical phenomena of heredity must be regarded as secondary to the physical or

---

[1] In saying this we make no assumption as to the particular cell-division at which differentiation occurs. This may be one of the maturation-divisions, or it may perhaps be much earlier.

4

mechanical phenomena by which the cells and their constituents are divided and separated. When therefore we speak of the *essential* phenomena of heredity we mean the mechanics of division, especially, though not, as we shall see, exclusively, of *cell*-division; and in the relation between the two halves of the dividing cell we have the problem presented in what seems to be its simplest form.

In attempting to form some conception of the processes by which bodily characteristics are transmitted, or—to avoid that confusing metaphor of "transmission"—how it comes about that the offspring can grow to resemble its parent, continuity of the germ-substance which in some animals is a visible phenomenon,[2] gives at least apparent help. An egg for example on becoming adult develops in certain parts a particular pigment. The eggs of that adult when they reach the appropriate age develop the same pigment. We have no clear picture of the mechanism by which this process is effected, but when we realise that the pigment results from the interaction of certain substances, and that since all the eggs are in reality pieces of the same material, it seems, unless we inquire closely, not unnatural that the several pieces of the material should exhibit the same colours at the same periods of their development. The continuity of the material of the germs suggests that there is a continuity of the materials from which the pigment is formed, and that thus an actual bit of those substances passess into each egg ready at the appropriate moment to generate the pigment. The argument thus outlined applies to all *substantive* characteristics. In each case we can imagine, if we will, the appearance of that characteristic as due to the contribution of its rudiment from the germ tissues.

When we consider more critically it becomes evident that the aid given by this mental picture is of very doubtful reality, for even if it were true that any predestined particle actually corresponding with the pigment-forming materials is definitely

---

[2] From the recent discoveries of Erwin Baur we are led to surmise that in the flowering plants the sub-epidermal layer, or some of its elements, may legitimately be regarded as a similar germ-substance, continuous in Weismann's sense.

passed on from germ to germ, yet the power of increase which must be attributed to it remains so incomprehensible that the mystery is hardly at all illuminated.

When however we pass from the substantive to the meristic characters, the conception that the character depends on the possession by the germ of a particle of a specific material becomes even less plausible. Hardly by any effort of imagination can we see any way by which the division of the vertebral column into x segments or into y segments, or of a Medusa into 4 segments or into 6, can be determined by the possession or by the want of a material particle. The distinction must surely be of a different order. If we are to look for a physical analogy at all we should rather be led to suppose that these differences in segmental numbers corresponded with changes in the amplitude or number of dividing waves than with any change in the substance or material divided.

PHENOMENA OF DIVISION

I have said that in the division of a cell we seem to see the problem in its simplest form, but it is important to observe that the problem of division may be presented by the bodies of animals and plants in forms which are independent of the divisions between cells. The existence of pattern implies a repetition of parts, and repetition of parts when developed in a material originally homogeneous can only be created by division. Cell-division is probably only a special case of a process similar to that by which the pattern of the skeleton is laid down in a unicellular body such as that of a Radiolarian or Foraminiferan. Attempts have lately been made to apply mathematical treatment to problems of biology. It has sometimes seemed to me that it is in the geometrical phenomena of life that the most hopeful field for the introduction of mathematics will be found. If anyone will compare one of our animal patterns, say hat of a zebra's hide, with patterns known to be of purely mechanical production, he will need no argument to convince him that there must be an essential similarity between the processes by which the two kinds of patterns were made and that parts at least of

the analysis applicable to the mechanical patterns are applicable
to the zebra stripes also. Patterns mechanically produced are
of many and very diverse kinds. One of the most familiar
examples, and one presenting some especially striking analogies
to organic patterns, is that provided by the ripples of a mackerel
sky, or those made in a flat sandy beach by the wind or the ebbing
tide. With a little search we can find among the ripple-marks, and
in other patterns produced by simple physical means, the closest
parallels to all the phenomena of striping as we see them in our
animals. The forking of the stripes, the differentiation of two
"faces," the deflections round the limbs and so forth, which in the
body we know to be phenomena of division, are common both to
the mechanical and the animal patterns. We cannot tell what in
the zebra corresponds to the wind or the flow of the current, but
we can perceive that in the distribution of the pigments, that is
to say, of the chromogen-substances or of the ferments which
act upon them, a rhythmical disturbance has been set up which
has produced the pattern we see; and I think we are entitled to
the inference that in the formation of patterns in animals and
plants mechanical forces are operating which ought to be, and
will prove to be, capable of mathematical analysis. The comparison between the striping of a living organism and the sand-
ripples will serve us yet a little farther, for a pattern may either
be formed by actual cell-divisions, and the distribution of differentiation coincidently determined, or—as visibly in the pigmentation of many animal and plant tissues—the pattern may
be laid down and the pigment (for example) distributed through
a tissue across or independently of the cell-divisions of the tissue.
Our tissues therefore are like a beach composed of sands of
different kinds, and different kinds of sands may show distinct
and interpenetrating ripples. When the essential analogy between these various classes of phenomena is perceived, no one
will be astonished at, or reluctant to admit, the reality of discontinuity in Variation, and if we are as far as ever from knowing
the actual causation of pattern we ought not to feel surprised that
it may arise suddenly or be suddenly modified in descent. Biologists have felt it easier to conceive the evolution of a striped

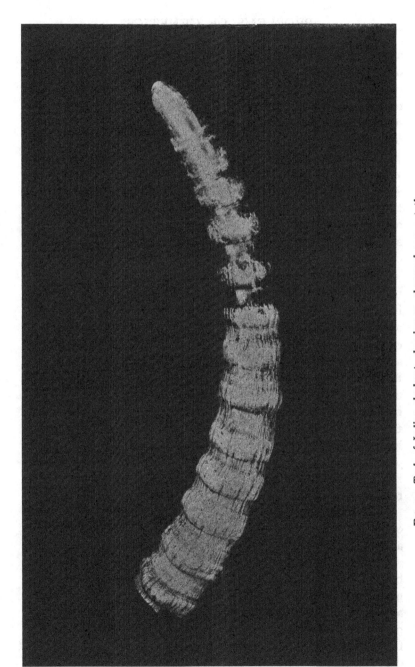

FIG. 1. Tusk of Indian elephant, showing an abnormal segmentation.

animal like a zebra from a self-coloured type like a horse (or
of the self-coloured from the striped) as a process involving many
intergradational steps; but so far as the *pattern* is concerned, the
change may have been decided by a single event, just as the
multitudinous and ordered rippling of a beach may be created
or obilterated at one tide.

This point is well illustrated by the tusk of an Indian elephant
which I lately found in a London sale-room. This tusk is by
some unknown cause, presumably a chronic inflammation,
thrown up into thirteen well-marked ridges which closely simulate
a series of segments (Fig. 1). Whatever the cause the condition
shows how easily a normally unsegmented structure may be
converted into a series of repeated parts.

The spread of segmentation through tissues normally unsegmented is very clearly exemplified in the skates' jaws shown in
in Fig. 2. The right side of the upper figure shows the normal
arrangement in the species *Rhinoptera jussieui*, but the structure
on the left side is very different. The probable relations of the
several rows of teeth to the normal rows is indicated by the lettering, but it is evident that by the appearance of new planes
of division constituting separate centers of growth, the series has
been recast. The pattern of the left side is so definite that had
the variation affected the right side also, no systematist would
have hesitated to give the specimen a new specific name. The
other two drawings show similar variations of a less extensive
kind, the nature of which is explained by the lettering of the
rows of teeth.

This power to divide is a fundamental attribute of life, and
of that power cell-division is a special example. In regard to
almost all the chief vital phenomena we can say with truth that
science has made some progress. If I mention respiration, meta-
bolism, digestion, each of these words calls to mind something
more than a bare statement that such acts are performed by an
animal or a plant. Each stands for volumes of successful experiment and research, But the expression cell-division, the
fundamental act which typifies the rest, and on which they all
depend, remains a bare name. We can see with the microscope

the outward symptoms of division, but we have no surmise as to the nature of the process by which the division is begun or accomplished. I know nothing which to a man well frained in scientific knowledge and method brings so vivid a realisation

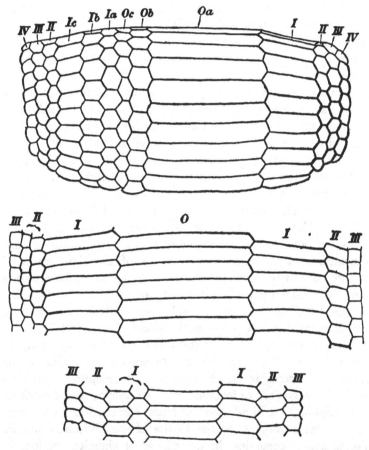

FIG. 2. Jaws of Skates (*Rhinoptera*) showing meristic variation. (For a detailed discussion see *Materials for the Study of Variation*, p. 259.)

of our ignorance of the nature of life as the mystery of cell-division. What is a living thing? The best answer in few words that I know is one which my old teacher, Michael Foster, used to give in his lectures introductory to biology. "A living thing is a vortex of chemical and molecular change." This description gives much, if not all, that is of the essence of life. The living thing is unlike ordinary matter in the fact that, through it, matter is always passing. Matter is essential to it; but, provided that the flow in and out is unimpeded, the life-process can go on so far as we know indefinitely. Yet the living "vortex" differs from all others in the fact that it can divide and throw off other "vortices," through which again matter continually swirls.

We may perhaps take the parallel a stage further. A simple vortex, like a smoke-ring, if projected in a suitable way will twist and form two rings. If each loop as it is formed could grow and then twist again to form more loops, we should have a model representing several of the essential features of living things.

It is this power of spontaneous division which most sharply distinguishes the living from the non-living. In the excellent book dealing with the problems of development, lately published by Mr. Jenkinson a special emphasis is very properly laid on the distinction between the processes of division, and those of differentiation. Too often in discussions of the developmental processes the distinction is obscured. He regards differentiation as the "central difficulty." "Growth and division of the nucleus and the cells," he tells us, are side-issues. This view is quite defensible, but I suspect that the division *is* the central difficulty, and that if we could get a rationale of what is happening in cell-division we should not be long before we had a clue to the nature of differentiation. It may be self-deception, but I do not feel it impossible to form some hypothesis as to the mode of differentiation, but in no mood of freest speculation are we ever able to form a guess as to the nature of the division. We see differentiations occurring in the course of chemical action, in some phenomena of vibration and so forth: but where do we see anything like the spontaneous division of the living cell? Excite a gold-leaf electroscope, and the leaves separate, but we know that is because they were double before. In electrolysis various substances separate out at the positive and negative poles respectively. Now if in cell-division the two daughter-cells were always dissimilar—that is to say, if differentiation always occurred—we could conceive some rough comparison with such dissociations. But we know the dissimilarity between daughter-cells is not essential. In the reproduction of unicellular organisms and many other cases, the products formed at the two poles are, so far as we can tell, identical. Any assumption to the contrary, if we were disposed to make it, would involve us in difficulties still more serious. At any rate, therefore, if differentiation be really the central difficulty in development, it is division which is the essential problem of heredity.

Sir George Darwin and Professor Jeans tell us that "gravitational instability" consequent on the condensation of gases is "the primary agent at work in the actual evolution of the universe," which has led to the division of the heavenly bodies. The greatest advance I can conceive in biology would be the discovery of the nature of the instability which leads to the c ntinual division of the cell. When I look at a dividing cell I feel as an astronomer might do if he beheld the formation of a double star: that an original act of creation is taking place before me. Enigmatical as the phenomenon seems, I am not without hope that, if it were studied for its own sake, dissociated from the complications which obscure it when regarded as a mere incident in development, some hint as to the nature of division could be found. It is I fear a problem rather for the physicist than for the biologist. The sentiment may not be a popular one to utter before an assembly of biologists, but looking at the truth impersonally I suspect that when at length minds of first rate analytical power are attracted to biological problems, some advance will be made of the kind which we are awaiting.

The study of the phenomena of bodily symmetry offers perhaps the most hopeful point of attack. The essential fact in reproduction is cell-division, and the essential basis of hereditary resemblance is the symmetry of cell-division. The phenomena of twinning provide a convincing demonstration that this is so. By twinning we mean the production of equivalent structures by division. The process is one which may affect the whole body of an animal or plant, or certain of its parts. The term twin as ordinarily used refers to the simultaneous birth of two individuals. Those who are naturalists know that such twins are of two kinds, (1) twins that are not more alike than any other two members of the same family, and (2) twins that are so much alike that even intimate friends mistake them. These latter twins, except in imaginative literature, are always of the same sex.

It is scarcely necessary for me to repeat the evidence from which it has been concluded that without doubt such twins arise by division of the same fertilised ovum. There is a perfect series of gradations connecting them with the various forms of double monsters united by homologous parts. They have been shown several times to be enclosed in the same chorion, and the proofs of experimental embryology show that in several animals by the separation of the two first hemispheres of a dividing egg twins can be produced. Lastly we have recently had the extraordinarily interesting demonstration of Loeb, to which I may specially refer. Herbst some years ago found that in sea water, from which all lime salts had been removed, the segments of the living egg fall apart as they are formed. Using this method Loeb has shown that a temporary immersion in lime-free sea water may result in the production of 90 per cent. of twins. We are therefore safe in regarding the homologous or "identical" twins as resulting fro the divisions of one fertilised egg, while the non-identical or "fraternal" twins, as they are called, arise by the fertilisation of two separate ova.[3]

In the resemblance of identical twins we have an extreme case

[3] These fraternal twins, which show no special resemblance to each other, are like the multiple births of other animals, and there is no disposition for them to be of the same sex. In the sheep, for example, statistics show that the frequency of pairs of twins, male and female, is approximately double that of the frequency of pairs, both male or both female, as it should be if the sex-distribution were fortuitous. For instance Bernadin (*La Bergerie de Rambouillet*, 1890, p. 100) gives the following figures for twin-lambs in Merinos: both male, 87; both female, 83; sexes mixed, 187. The 9-banded Armadillo (*Dasypus novemcinctus*), in which the young born in one litter are said to be always of one sex, is the only known exception in Vertebrates, and is presumably a genuine case of normal polyembryony (see especially, Rosner, *Bull. Ac. Soc. Cracovie*, 1901, p. 443, and Newman and Patterson, *Biol. Bull.*, XVII, 1909, p. 181, and an important paper lately published by H. H. Newman and J. T. Patterson, *Jour. Morph.*, 1911, XXII, p. 855.

of hereditary likeness[4] and a proof, if any were needed, that the cause of individual variation is to be sought in the differentiation of germ-cells. The resemblance of identical twins depends on two circumstances, First, since only two germ-cells take part in their production, difference between the germ cells of the same individual cannot affect them. Secondly the division of the fertilised ovum, the process by which they became two instead of one, must have been a symmetrical division. The structure of twins raises however one extremely significant difficulty, which as yet we cannot in any way explain. The resemblance between twins is a phenomenon of symmetry, like the resemblance between the two sides of a bilaterally symmetrical body. Not only is the general resemblance readily so interpreted, but we know also that in double monsters, namely unseparated twins, various anatomical abnormalities shown by the one half-body are frequently shown by the other half-also.[5] The two belong to one system of symmetry How then does it happen that the body of one of a pair of twins does not show a transposition of viscera? We know that the relation of right and left implies that the one should be the mirror-image of the other. Such a relation of images may be maintained even in minute details. For example if the same pattern of fingerprint is given by the fingers of the two hands, one is the reverse of the other. In double monsters, namely unseparated twins, there is evidence that an inversion of viscera does occur with some frequency. Evidence from such cases is not so clear and simple as might be expected, because as a matter of fact, the heart and stomach, upon which the asymmetry of the viscera chiefly depend, are usually common to the two bodies. Duplicity generally affects either the anterior end alone, or the posterior end alone. The division is generally *from the heart forwards*, giving two heads and two pairs of anterior limbs on a common trunk, or *from the heart backwards*, giving two pairs of posterior limbs with the anterior body common. In either case, though

---

[4] A good collection of evidence as to disease in homologous twins was lately published by E. A. Cockayne, *Brit. Jour. Child. Diseases*, Nov., 1911.

[5] Cp. Windle, B. C. A., *Jour. Anat. Phys.*, XXVI, p. 295.

the bodies may be grouped in a common system of symmetry, neither can be proved to show definite reversal of the parts. To see that reversal recourse must be had to more extreme duplications, such as the famous Siamese Twins. They, as a matter of fact, were an excellent instance of the proposition that twins are related as mirror-images, for both of them had eleven pairs of ribs instead of the normal twelve, and one of them had a partial reversal of viscera.[6] (Küchenmeister, *Verlagerung*, etc., p. 204.)

If anyone could show how it is that neither of a pair of twins has transposition of viscera the whole mystery of division would, I expect, be greatly illuminated.[7] At present we have simply to accept the fact that twins, by virtue of their detachment from each other, have the power of resuming the polarity which is proper to any normal individual. It was nevertheless with great interest that I read Wilder's recent observation[8] that occasionally in identical twins the finger-print of one or both the index-fingers may be reversed, showing that there is after all some truth in the notion that reversal should occur in them.

There is another phenomenon by twinning which, if we could understand it, might help. I refer to the free-martin, the subject of one of John Hunter's masterpieces of anatomical description. In horned cattle twin births are rare, and when twins of opposite sexes are born, the male is perfect and normal, but the repro-

[6] Mr. E. Nettleship tells me that in the course of collecting pedigrees of families containing colour-blind members he has discovered two cases (shortly to be published) of pairs of twins, which on account of their very close resemblances must be deemed homologous, one of each pair being colour-blind and the other normal. Such a distinction between closely similar twins is most curious and unexpected.

[7] Another paradoxical phenomenon of the same nature occurs in the Narwhal The males normally have the *left* tusk alone developed, the corresponding right tusk remaining as an undeveloped rudiment in its socket. The left tusk is a left-handed screw. Occasionally the right tusk is also developed and grows to the same length as that of the left side, but in such specimens the right tusk is also a left-hand screw like the tusk of the other side, instead of being reversed as we should certainly have expected. It need scarcely be remarked that in the case of the horns of antelopes, and in other examples of spiral organs arranged in pairs, that of one side of the body is the mirror image of that on the other side. The Narwhal's tusks in being both twisted in the same direction are thus highly anomalous, and are comparable with pairs of twins.

[8] Wilder, H. H., *Amer. Jour. Anat.*, 1904, III, p. 452.

ductive organs of the female are deformed and sterile, being known as a free-martin. The same thing occasionally happens in sheep, suggesting that in sheep also twins may be formed by the division of one ovum; for it is impossible to suppose that mere development in juxtaposition can produce a change of this character. I mention the free-martin because it raises a question of absorbing interest. It is conceivable that we should interpret it by reference to the phenomenon of gynandromorphism, seen occasionally in insects, and also in birds as a great rarity. In the gynandromorph one side of the body is male, the other female. A bullfinch for instance has been described with a sharp line of division down the breast between the red feathers of the cock on one side and the brown feathers of the hen on the other. (Poll, H., *SB. Ges. Nat. Fr.*, Berlin, 1909, p. 338.) In such cases neither side is sexually perfect. If the halves of such a gynandromorph came apart, perhaps one would be a free-martin.

The behaviour of homologous twinning in heredity has been little studied. It does not exist as a normal feature in any animal which is amenable to experiment, and we cannot positively assert that a comparable phenomenon exists in plants; for in them—the Orange, for example—polyembryony may evidently be produced by a parthenogenetic development of nucellar tissue. It is possible that in Man twinning is due to a peculiarity of the mother, not of the father. It may and not rarely does descend from mother to daughter, but whether it can be passed on through a male generation to a daughter again, there is not sufficient evidence to show. The facts as far as they go are consistent with the inference which may be drawn from Loeb's experiment, that the twinning of a fertilized ovum may be determined not by the germ-cells which united to form it, but by the environment in which it begins to develop. The opinion that twinning may descend through the male directly has been lately expressed by Dr. J. Oliver in the *Eugenics Review* (1912), on the evidence of cases in which twins had occurred among the relations of fathers of twins, but I do not know of any comprehensive collection of evidence bearing on the subject.

Besides twinning of the whole body a comparable duplicity

of various parts of the same body may occur. Such divisions affect especially those organs which have an axis of bilateral symmetry, such as the thumb, a cotyledon, a median petal, the frond of a fern or the anal fin of a fish. From the little yet known it is clear that the genetic analysis of these conditions must be very difficult, but evidence of any kind regarding them will be valuable. We want especially to know whether these divisions are due to the *addition* of some factor or power which enables the part to divide, or whether the division results from the *absence* of something which in the normal body prevents the part from dividing. Breeding experiments, so far as they go, suggest that the less divided state is usually dominant to the more divided.[9] The two-celled Tomato fruit is dominant to the many-celled type. The Manx Cat's tail, with its suppression of caudal segmentation is a partial dominant over the normal tail. The tail of the Fowl in what is called the "Rumpless" condition is at least superficially comparable with that of the Manx Cat, and though the evidence is not wholly consistent, Davenport obtained facts indicating that this suppressed condition of the caudal vertebrae is an imperfect dominant.[10]

Some evidence may also be derived from other examples of differences which at first sight appear to be substantive though they are more probably meristic in ultimate nature. The distinction between the normal and the "Angora" hair of the Rabbit is a case in point. We can scarcely doubt that one of the essential differences between these two types is that in the Angora coat the hair-follicles are more finely divided than they are in the normal coat, and we know that the normal, or less-divided condition, is dominant to the Angora, or more finely divided.

In the case of the solid-hoofed or "mule-footed" swine, the

[9] Polydactylism which is often a dominant and the web-foot of Pigeons which is recessive should be remembered as possible exceptions (see p. 49).

[10] Davenport inclined at first to regard rumplessness as a recessive, but in his latest publication on the subject he definitely concludes that it is an imperfect dominant. This conclusion accords well with evidence quoted by Darwin (*An. and Plts.*, II, ed. 2, p. 4) that rumpless fowls may throw tailed offspring. (*Amer. Nat.*, 1910, XLIV, p. 134.)

evidence shows, as Spillman has lately pointed out,[11] that the condition behaves as a dominant. The essential feature of this abnormality is that the digits III and IV are partially united. The union is greatest peripherally. Sometimes the

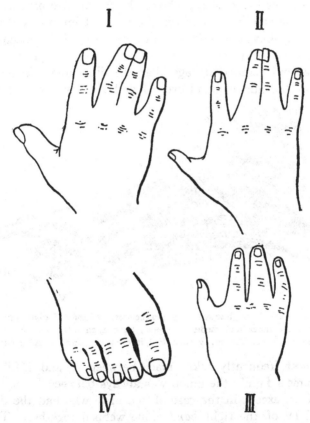

FIG. 3. *I, II, III,* various degrees of syndactyly affecting the medius and annularis in the hand; *IV,* syndactyly affecting the index and medius in the foot. (After Annandale.)

third phalanges only are joined to form one bone, but the second and even the first phalanges may also be compounded together. Here the variation is obviously meristic and consists in a failure

[11] Spillman, W. J., *Amer. Breeders Mag.,* 1910, I, p. 178.

to divide, the normal separation of the median digits of the foot being suppressed.

Webbing between the digits, in at least some of its manifestations, is a variation of similar nature. The family recorded by Newsholme[12] very clearly shows the dominance of this condition. The case is morphologically of great interest and must undoubtedly have a bearing on the problems of the mechanics of Division. In discussing the phenomena of syndactylism I pointed out some years ago that the digits most frequently united in the human hand are III and IV, while in the foot,

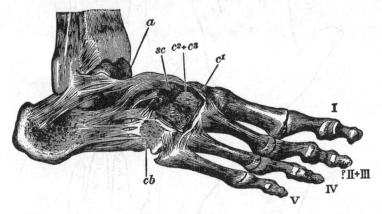

FIG. 4. Case of complete syndactyly in the foot. *II* and *III*, digit apparently representing the index and medius. $c^2 + c^3$, bone apparently representing the middle and external cuneiform; *cb*, cuboid; c^1, internal cuneiform. (After Gruber.)

union most frequently takes place between II and III.[13] In Newsholme's family the union was always between II and III of the foot, except in the case of one male who had the digits III and IV of the right *hand* alone webbed together. There can be little doubt that the geometrical system on which the foot is planned has an axis of symmetry passing between the digits II and III, while the corresponding axis in the hand passes between III and IV. Union between such digits may therefore be regarded as comparable with any non-division or "coalescence" of lateral structures in a middle line, and when as in these ex-

[12] Newsholme, *Lancet*, December 10, 1910, p. 1690.
[13] *Materials for the Study of Variation*, 1894, p. 358.

amples such a condition is shown to be a dominant we cannot avoid the inference that some concrete factor has the power of suppressing or inhibiting this division. Figs. 3 and 4 illustrate degrees of union between digits in the human hand and foot.

It is not in question that various other forms of irregular webbing and coalescence of digits exist, and respecting the genetic behaviour of these practically nothing is as yet known, Such a case is described by Walker,[14] in which the first and second metacarpals of both feet were fused in mother and daughter, and several more are found in literature. Contrasted with these phenomena we have the curious fact that in the Pigeon, Staples-Browne found webbing of the toes a *recessive* character. The question thus arises whether this webbing is of the same nature as that shown to be a dominant in Man, and indeed whether the phenomenon in pigeons is really meristic at all. There is some difference perceptible between the two conditions; for in Man there is not so much a development of a special web-like skin uniting the digits as a want of proper division between the digits themselves, and in extreme cases two digits may be represented by a single one. In the Pigeon I am not aware that a real union of this kind has ever been observed, and though the web-like skin may extend the whole length of the digits and be so narrow as to prevent the spread of the toes, it may, I think, be maintained that the unity of the digits is unimpaired. For the present the nature of this variation in the pigeon's feet must be regarded as doubtful, and we should note that if it is actually an example of a more perfect division being dominant to a less perfect division, the case is a marked exception to the general rule that non-division is dominant to division.

Reference must also be made to the phenomenon of fasciation in the stems of plants. As Mendel showed in the case of *Pisum* this condition is often a recessive. The appearances suggest that the difference between a normal and a fasciated plant consists in the inability of the fasciated plant to separate its lateral branches. The nature of the condition is however very

[14] Walker, G., *Johns Hopkins Hospital Bulletin*, XII, 1901, p. 129.

obscure and it is equally likely that some multiplication of the growing point is the essential phenomenon.[15]

Stockard's interesting experiments[16] illustrate this question. He showed that by treating the embryos of a fish (*Fundulus heteroclitus*) with a dilute solution of magnesium salts, various cyclopian monstrosities were frequently produced. These have been called cases of *fusion* of the optic vesicles. I would prefer to regard them as cases of a division suppressed or restricted by the control of the environment. Conversely, the splendid discovery of Loeb, that an unfertilised egg will divide and develop parthenogenetically without fertilisation, as a consequence of exposure to various media, may be interpreted as suggesting that the action of those media releases the strains already present in the ovum, though I admit that an interpretation based on the converse hypothesis, that the medium acts as a stimulus, is as yet by no means excluded.

In these cases we come nearest to the direct causation or the direct inhibition of a division, but the meaning of the evidence is still ambiguous. I incline to compare Loeb's parthenogenesis with the development (and of course accompanying cell-division) of dormant buds on stems which have been cut back.

It is interesting to note that sometimes as an abnormality, the faculty of division gets out of hand and runs a course apparently uncontrolled. A remarkable instance of this condition is seen in *Begonia* "*phyllomaniaca*," which breaks out into buds at any point on the stem, petioles, or leaves, each bud having, like other buds, the power of becoming a new plant if removed. We would give much to know the genetic properties of *B. phyllomaniaca*, and in conjunction with Mr. W. O. Backhouse I have for some time been experimenting with this plant. It proved totally sterile. Its own anthers produce no pollen, and all attempts to fertilise it with other species failed though the pollen of a great number of forms was tried.

Recently however we have succeeded in making plants which

---

[15] Cp. R. H. Compton, *New Phytologist*, 1911, p. 249.
[16] *Arch. f. Entwickelungsmech.*, 1907, XXIII, p. 249.

are in every respect *Begonia phyllomaniaca*, so far as the characters of stems and leaves are concerned. These plants, of which we have sixteen, were made by fertilising *B. heracleifolia* with *B. polyantha*. They are all beginning to break out in "phyllomania." As yet they have not flowered, but as they agree in all details with *phyllomaniaca* there can be little doubt that the original plant bearing that name was a hybrid similarly produced. The production of "phyllomania" on a hybrid Begonia has also been previously recorded by Duchartre.[17] In this case the cross was made between *B. incarnata* and *lucida*. The synonymy of the last species is unfortunately obscure, and I have not succeeded in repeating the experiment.

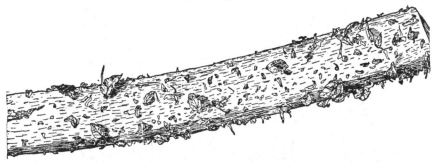

FIG. 5. Piece of petiole of *Begonia phyllomaniaca*. The proximal end is to the right of the figure.

From these facts it seems practically certain that the condition is one which is due to the meeting of complementary factors. At first sight we may incline to think that the phyllomania is in some way due to the sterility. This however cannot be seriously maintained; for not only is sterility in plants not usually associated with such manifestations, but we know a Begonia called "Wilhelma" which is exactly *phyllomaniaca* and equally sterile, though it has no trace of phyllomania. This plant arose in the nurseries of MM. P. Bruant of Poitiers, and has generally been described as a seedling of *phyllomaniaca*, but from the total sterility of that form this account of its origin must be set aside.

The phenomenon in this case can hardly be regarded as

[17] Bull. Soc. Bot. de France, xxxiv, 1887, p. 182.

due to the excitation of dormant buds, for it is apparent on examination that the new growths are not placed in any fixed geometrical relation to the original plant. They arise on the

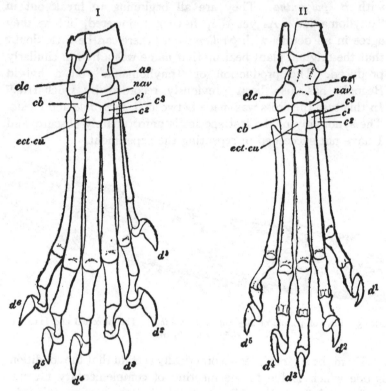

FIG. 6. Two right hind feet of polydactyle cats. *II* shows the lowest development of the condition yet recorded. The digit, d_1, which stands as hallux is fully formed and has three phalanges. Both it and the digit marked d_2 are formed as *left* digits. In the normal hind foot of the cat the hallux is represented by a rudiment only.

I shows a further development of the condition. In this foot there are *six* digits. d_1 has two phalanges, but both it and d^2 and d^3 are shaped as left digits. Thus d^3, which in the normal foot would be shaped as a right digit, is transformed so as to look like a *left* digit.

petiole, for example, as small green outgrowths each of which gradually becomes a tiny leaf. The attitude of these leaves is quite indeterminate, and they may point in any direction, some having their apices turned peripherally, some centrally, and others in various oblique or transverse positions (Fig. 5). These little leaves are thus comparable with seedlings, in that their polarity is not related to, or consequent upon that of the parent plant. They have in fact that "individuality," which we associate with germinal reproduction.

There are many curious phenomena seen in the behaviour of parts normally repeated in bilateral symmetry which may some day guide us towards an understanding of the mechanics of division. A part like a hand, which needs the other hand to complete its symmetry, cannot twin by mere division, yet by proliferation and special modifications on the radial side of the same limb, even a hand may be twinned. In the well known polydactyle cats a change of this kind is very common and indeed almost the rule. When extra digits appear at the inner (tibial) side of the limb, they are shaped as digits of the other side, and even the normal digit II (index) is usually converted into the mirror-image of its normal self. The limb then develops a new symmetry in itself. Nevertheless it is not easy to interpret these facts as meaning that there has been some interruption in the control which one side of the body exercises over the other. The heredity of polydactylism is complex but there is little doubt that the condition familiar in the Cat is a dominant. In some human cases also the descent is that of a dominant, but irregularities are so frequent that no general rule can yet be perceived. The dominance of such a condition is an exception to the principle that the less-divided is usually dominant to the more-divided, a fact which probably should be interpreted as meaning that divisions are of more than one kind.

Among ordinary somatic divisions, whether of organs, cells, or patterns of differentiation, the control of symmetry is usually manifested. There is however one class of somatic differentiations which are exceptionally interesting from the fact that they may show a complete independence of such geometrical control. The most familiar examples of these geometrically uncontrolled Variations are to be seen in bud-sports. The normal differentiation of the organs of a plant is arranged on a definite geometrical system, which to those who have never given special attention to such things before, will often seem surprisingly precise. The arrangement of the leaves on uninjured, free-growing shoots can generally be seen to follow a very definite order, just as do the flowers or the parts of the flowers. If however bud sports occur, then though the parts included in the sports show all the geometrical peculiarities proper to the sport-variety, yet the sporting-buds themselves are not related to each other according to any geometrical plan.

A very familiar illustration is provided by the distribution of colour in those Carnations that are not self-coloured. The pigment may, as in Picotees, be distributed peripherally with great regularity to the edges of the petals; or, as in Bizarres and Flakes, it may be scattered in radial sectors which show no geometrical regularity. Now in this case the pigments are the same in both types of flower, and the chemical factors concerned in their production must surely be the same. The difference must lie in the mechanical processes of distribution of the pigment. In the Picotee we see the orderly differentiation which we associate with normality; in the Bizarre we see the disorderly differentiation characteristic of bud-sports. The distribution of colour in this case lies outside the scheme of symmetry of the plant.

Such a distribution is characteristic of bud-sports, and of certain other differentiations in both plants and animals, which I cannot on this occasion discuss. Now reflexion will show that these facts have an intimate bearing on the mechanical problems of heredity. For first in the bud-sports we are witnessing the distribution of factors which distinguish genetic varieties. We do not know the physical nature of those factors, but if we must give them a name, I suppose we should call them "ferments" exactly as Boyle did in 1666. He is discussing how it comes about that a bud, budded on a stock, becomes a branch bearing the fruit of its special kind. He notes that though the bud inserted be "not so big oftentimes as a Pea," yet "whether by the help of some peculiar kind of Strainer or by the Operation of some powerful Ferment lodged in it, or by both these, or some other cause," the sap is "so far changed as to constitute a Fruit quite otherwise qualify'd." [18] We can add nothing to his speculation, and we believe still that by a differential distribution of "ferments" the sports are produced. All the factors are together present in the normal parts; some are left out in the sport. In an analogous case however, that of a variegated *Pelargonium* which has green and also albino shoots, Baur proved that the shoots pure in colour are also pure in their posterity. There can be no doubt that the sports of Carnations, Azaleas, Chrysanthemums, etc., would behave in the same way.

The well-known Azaleas Perle de Ledeburg, President Kerchove, and *Vervaeana* are familiar illustrations. Perle de Ledeburg is predominantly white, but it has red streaks in some of its flowers. It not very rarely gives off a self-red sport. This is evidently due to the development of a bud in a red-bearing area of the stem. The red in this plant is not under "geometrical control." Many plants have white flowers with no markings, but if the red markings are geometrically ordered differentiations, no self-coloured sports are formed. The case of *Vervaeana* is a good illustration of this proposition. It has white flowers with red markings arranged in an orderly manner on the lower parts of the petals, especially on the dorsal petals. This is one of the Azaleas most liable to have red sports, and at first sight it might seem that the sport represented the red of the central marks. Examination however of a good many flowers shows that irregular red streaks like those of Perle de Ledeburg occur, about as commonly as in that variety. *Vervaeana* in fact is Perle de Ledeburg with *definite* red markings added, and its red sports obviously are those branches the germs of which came in a patch of the stem bearing these red elements. That this is the true account is rendered quite obvious by the fact that the red of the sport is a colour somewhat different from that of the definite marks, and that these marks are still present on the red ground of the sporting flowers.

It will be understood that these remarks apply to those cases in which the production of sports is habitual or frequent, and

[18] R. Boyle, *The Origine of Formes and Qualities*, Oxford, 1666.

I imagine in all such examples it will be found that there are indications of irregularity in the distribution of the differentiations such as to justify the view that they are not under that geometrical control which governs the normal differentiation of the parts. The question next arises whether these considerations apply also to the production of a bud-sport as a rare exception, but by the nature of the case it is not possible to say positively whether the appearance of an exceptional sport is due to the unsuspected presence of a pre-existing fragment of material having a special constitution, or to the origin, *de novo*, of such a material. For instance one of the garden forms of *Pelargonium* known as *altum* is liable perhaps once in some hundreds of flowers to have one or two magenta petals. The normal colour is a brilliant red; and as we may be fairly sure that this red is recessive to magenta the interpretation would be quite different according as the appearance of the magenta is regarded as due to the presence of small areas endowed with magentaness, or to the spontaneous generation of the factor for that pigment. Either interpretation is possible on the facts, but the view that the whole plant has in it scarce mosaic particles of magenta seems on the whole more consistent with present knowledge.

In *Pelargonium altum* the enzyme causing the magenta colours must be distributed in very small areas, but a case in which the magenta is similarly arranged in a much coarser patchwork may be seen in the *Pelargonium* " Don Juan," which often bears whole trusses or branches of red flowers upon plants having the normal dominant magenta trusses. In most cases there is little doubt that though the magenta flowered parts can " sport" to red, the red parts could not produce the magenta flowers.

The asymmetrical, or to speak more precisely, the disorderly, mingling of the colours in the somatic parts is thus an indication of a similarly disorderly mixing of the factors for those colours in the germ-tissues, so that some of the gametes bear enough of the colour-factors to make a self-coloured plant, while others bear so little that the plant to which they give rise is a patchwork. If this view is correct we may extend it so far as to con-

sider whether the fineness or coarseness of the mixture visible
in the flowers or leaves may not give an indication of the degree
to which the factors are subdivided among the germ-cells. We
know very little about the genetic properties of striped varieties.
In both *Antirrhinum* and *Mirabilis* it has been found that the
striped may occasionally and irregularly throw self-coloured
plants, and therefore the striping cannot be regarded simply as
a recessive character. On the other hand in *Primula Sinensis*
there are well-known flaked varieties which ordinarily at least
breed true. Whether these ever throw selfs I do not know,
but if they do it must be quite exceptionally. The power of
these flaked plants to breed true is, I suspect, connected with
the fact that in their flowers the coloured and white parts are
intimately mixed, this intimate mixture thus being an indication
of a similarly intim te mixture in the germ-cells. It would be
important to ascertain whether self-fertilised seed from the occasional flowers in which the colour has run together to join a
large patch gives more self-coloured plants than the intimately
flaked flowers do.

The next fact may eventually prove of great importance.
We have seen that in bud-sports the differentiation is of the same
nature as that between pure types, and also that in the sporting
plant this differentiation is distributed without any reference
to the plant's axis, or any other consideration of symmetry.
Now among the germ-cells of a Mendelian hybrid exactly such
characters are being distributed allelomorphically, and there
again we have strong evidence for believing that the distribution
obeys no pattern. For example, we can in the case of seeds still
in situ perceive how the characters were distributed among the
germ-cells, and there is certainly no obvious pattern connecting
them, nor can we suppose that there is an actual pattern obscured.

Of this one illustration is especially curious. Individual
plants of the same species are, as regards the decussations of
their leaves and in other respects, *either rights or lefts*. The fact
is not emphasized in modern botany and is in some danger of
being forgotten. When, as in the flowers of Arum, some *Gladioli*,
Exacum, *St. Paulia*, or the fruits of *Loasa*, rights and lefts occur

on the same stem, they come off alternately. But if, as in the seedlings of Barley the twist of the first leaf be examined, it will be seen to be either a right- or left-handed screw. An ear of barley, say a two-row barley, is a definitely symmetrical structure. The seeds stand in their envelopes back to back in definite positions. Each has its organs placed in perfectly definite places. *If these seeds were buds* their differentiations would be grouped into a common plan. One might expect that the differentiations of these embryos would still fall into the pattern; but they do not, and so far as I have tested them, any one may be a right or a left, just as each may carry any of the Mendelian allelomorphs possessed by the parent plant, without reference to the differentiation of any other seed. The fertilisation may be responsible, but our experience of the allelomorphic characters suggest that the irregularity is in the egg-cells themselves.[19]

Germ cells thus differ from somatic cells in the fact that their differentiations are outside the geometrical order which governs the differentiation of the somatic cells. I can think of possible exceptions, but I have confidence that the rule is true and I regard it as of great significance.

The old riddle, what is an individual, finds at least a partial solution in the reply that an individual is a group of parts differentiated in a geometrically interdependent order. With the germ-cell a new geometrical order, with independent polarity is almost if not quite always, begun, and with this geometrical independence the power of rejuvenescence may possibly be associated.

The problems thus raised are unsolved, but they do not look insoluble. The solution may be nearer than we have thought. In a study of the geometry of differentiation, germinal and somatic, there is a way of watching and perhaps analyzing what may be distinguished as the mechanical phenomena of heredity. If any one could in the cases of the Picotee and the Bizarre Carnation, respectively, detect the real distinction between the two

[19] Remarkable experiments on this question have lately been carried out by R. H. Compton (*Camb. Phil. Soc.*, XV, 1910, p. 495), showing that in a certain Barley, "Plumage Corn," the average ratio of left to right is about 1.5. A fuller paper has since been published by Compton, *Jour. Genetics*, 1912, II, 1, p. 53.

types of distribution, he would make a most notable advance. Any one acquainted with mechanical devices can construct a model which will reproduce some of these distinctions more or less faithfully. The point I would not lose sight of is that the analogy with such models must for a long way be a true and valuable guide. I trust that some one with the right intellectual equipment will endeavor to follow this guide; and I am sanguine enough to think that a comprehensive study of the geometrical phenomena of differentiation will suggest to a penetrative mind that critical experiment which may one day reveal the meaning of spontaneous division, the mystery through which lies the road, perhaps the most hopeful, to a knowledge of the nature of life.

CHAPTER III

SEGMENTATION, ORGANIC AND MECHANICAL

Models may be and often have been devised imitating some of the phenomena of division, but none of them have reproduced the peculiarity which characterises divisions of living tissues, that *the position of chemical differentiation* is *determined by those divisions*. For example, models of segmentation, whether radial or linear, may be made by the vibration of plates as in the familiar Chladni figures of the physical laboratory, or by the bowing of a tube dusted on the inside with lycopodium powder, and in various other ways. The sand or the powder will be heaped up in the nodes or regions of least movement, and the patterns thus formed reproduce many of the geometrical features of segmentation. But in the segmentations of living things the nodes and internodes, once determined by the dividing forces, would each become the seat of appropriate and distinct chemical processes leading to the differentiation of the parts, and the deposition of the bones, petals, spines, hairs, and other organs in relation to the meristic ground-plan. The "ripples" of meristic division not merely divide but differentiate, and when a "ripple" forks the result is not merely a division but a re-duplication of the organ through which the fork runs. An example illustrating such a consequence is that of the half-vertebrae of the Python. On the left side the vertebra is single (Fig. 7) and bears a single rib, but on the right side a division has occurred with the result that two half-vertebrae, each bearing a rib, are formed, one standing in succession to the other. We cannot, indeed, imagine any operation of physiological division carried out in such an organ as a vertebra, passing through a plane at right angles to the long axis of the body, which does not necessarily involve the further process of reduplication.

As the meristic system of distribution spreads through the body, chemical differentiations follow in its track, with segmentation and pattern as the visible result. Could we analyse these simultaneous phenomena and show how it is that the places of chemical differentiation are determined by the system of division, progress would then be rapid. It is here that all speculation fails.

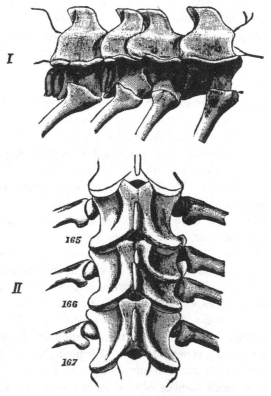

FIGS. 7 and 8. Two examples of imperfect division in the vertebræ of a python. *I*, the vertebræ 147–150 from the right side, showing imperfect division between the 148th and 149th. The condition on the left side of this vertebra was the same. *II*, the dorsal surface of vertebræ 165–167. On the right side the 166th is double and bears two ribs, but on the left side it is normal and has one rib only.

Many attempts have been made to interpret the processes of division and repetition, in terms of mechanics, or at least to refer them to their nearest mechanical analogies, so far with little success. The problem is beset with difficulties as yet insurmountable and of these one must be especially noticed. In the living thing the process by which repetition and patterns come into being consists partly in division but partly also in growth. We have no means of studying the phenomena of pattern-formation except in association with that of growth. Growth soon ceases unless division takes place, and if growth is impossible division soon ceases also. In consequence of this fact that the final pattern is partly a product of growth, it can never be used as unimpeachable evidence of the primary geometrical relations of the members as laid down in the divisions.

In the last chapter in referring to the problem of repetition I introduced an analogy, comparing the patterns of the organic world with those produced in unorganised materials by wave-motion. In the preliminary stage of ignorance, having no more trustworthy clue, I do not think it wholly unprofitable to consider the applicability of this analogy somewhat more fully. It possesses, as I hope to show, at least so much validity as to encourage the belief that morphology may safely discard one source of long-standing error and confusion.

Those who have studied the structure of parts repeated in series will have encountered the old morphological problem of "Serial Homology," which has absorbed so much of the attention of naturalists and especially of zoologists at various periods. This problem includes two separate questions. The first of these is the origin in evolution of the resemblance between two organs occurring in a repeated series, of which the fore and hind limbs of Vertebrates are the prerogative instance. From the fact that these resemblances can be traced very far, often into minute details of structure, many anatomists have inclined to the opinion that the resemblance must originally have been still more complete, and that the two limbs, for instance, must have acquired their present forms by the differentiation of two identical groups of parts.

Similar questions arise whenever parts are repeated in series, whether the series be linear or radial, and, though less obviously, even when the repetition is bilateral only. In each such example the question arises, is the resemblance between the parts the remains of a still closer resemblance, or is differentiation original? Sometimes the view that these parts have arisen by the differentiation of a series of identical parts is plausible enough, as for example when the peculiarities of various appendages of a Decapod Crustacean are referred to modifications of the Phyllopod series. In application to other cases however we soon meet with difficulty, and the suggestion that the segments of a vertebrate were originally all alike is seen at once to be absurd, for the reason that a creature so constituted could not exist, and that, differentiation of at least one anterior and one posterior segment, is an essential condition of a viable organism consisting of parts repeated in a linear series. Between these two terminal segments it is possible to imagine the addition of one segment, or of a series of approximately similar segments; but when once it is realised that the terminals must have been differentiated from the beginning, it will be seen that the problem of the origin of the resemblance between segments is not rendered more comprehensible by the suggestion that even the intervening members were originally alike. Seeing indeed that some differentiation must have existed primordially it is as easy to imagine that the original body was composed of a series grading from the condition of the anterior segment to that of the posterior, as any other arrangement. The existence of a linear or successive series in fact postulates a polarity of the whole, and in such a system the conception of an ideal segment containing all the parts represented in the others has manifestly no place. The introduction of that conception though sanctioned by the great masters of comparative anatomy, has, as I think, really delayed the progress of a rational study of the phenomena of division. The same notion has been applied to every class of repetition both in animals and plants, generally with the same unhappy results. In the cruder forms in which this doctrine was taught thirty years ago it is now seldom expressed, but modified presentations of it still survive and confuse our judgments.

The process of repetition of parts in the bodies of organisms is however a periodic phenomenon. This much, provided we remain free from prejudice as to the nature and causation of the period or rhythm, we may safely declare, and a comparison may thus be instituted between the consequences of meristic repetition

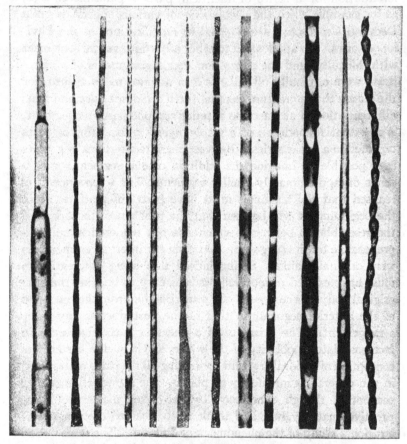

FIG. 9.   Osmotic growths simulating segmentation.   (After Leduc.)

in the bodies of living things and those repetitions which in the inorganic world are due to rhythmical processes. Of such processes there is a practically unlimited diversity and we have nothing to indicate with which of them our repetitions should rather be compared.

In some respects perhaps the best models of living organisms yet made are the "osmotic growths" produced by Leduc.[1] These curious structures were formed by placing a fragment of a salt, for instance calcium chloride, in a solution of some colloidal substance. As the solid takes up water from the solution a permeable pellicle or membrane is formed around it. The vesicle thus enclosed grows by further absorption of water, often extending in a linear direction, and in many examples this growth occurs by a series of rhythmically interrupted extensions. Some of the growths thus formed are remarkably like organic structures, and might pass for a series of antennary segments or many other organs consisting of a linear series of repeated parts. In admitting the essential resemblance between these "osmotic growths" and living bodies or their organs I lay less stress on the general conformation of the growths, which often as Leduc points out, recall the forms of fungi or hydroids, but rather on the fact that the interruptions in the development of these systems are so closely analogous to the segmentations or repetitions of parts characteristic of living things (Fig. 9). In the same way I am less impressed by Leduc's models of Karyokinesis, wonderful as they nevertheless are, for the division is here imitated by putting separate drops on the gelatine film. What we most want to know is how in the living creature one drop becomes two. The models of linear segmentation have the remarkable merit that they do in some measure imitate the process of actual division or repetition. So in a somewhat modified method Leduc, by causing the diffusion of a solution in a gelatine film, produced rhythmical or periodic precipitations strikingly reminiscent of various organic tissues, for here also the process of periodic repetition is imitated with success.

It is a feature common to these and to all other rhythmical repetitions produced by purely mechanical forces that there is resemblance between the members of the series, and that this similarity of conformation may be maintained in most complex detail. When however in the mechanical series some of the members differ from the rest we have no difficulty in recognising

---

[1] Stéphane Leduc, *Théorie Physico-Chymique de la Vie*, Paris, 1910.

that these differences—which correspond with the differentiations of the organic series—are due to special heterogeneity in
the conditions or in the materials, and it never occurs to us to
suppose that all the members must have been primordially alike.
For example, in the case of ripple-marks on the sand, which I
choose as one of the most familiar and obvious illustrations
of a repeated series due to mechanical agencies, if we notice
one ripple different in form from those adjacent to it, we do
not suppose that this variation must have been brought about by
deformation of a ripple which was at first formed like the others,
but we ascribe it to a difference in the sand at that point, or to a
difference in the way in which the wind or the tide dealt with it.
We may press the analogy further by observing that in as much
as such a series of waves has a beginning and an end, it possesses
polarity like that of the various linear series of parts in organisms,
and even the formation of each member must influence the
shape of its successor. Since in an organism the beginning and
end of the series are always included, some differentiation among
the repetitions must be inevitable. If therefore it be conceded,
as I think it must, that segmentation and pattern are the consequence of a periodic process we realize that it is at least as
easy to imagine the formation of such a series of parts having
family likeness combined with differentiation as it would be to
conceive of their arising primordially as a series of identical repetitions. The suggestion that the likenesses which we now perceive are the remains of a still more complete resemblance
is a substitution of a more complex conception for a simpler one.
    The other question raised by the problem of Serial Homology
is how far there is a correspondence between individual members
of series when the series differ from each other either in the
number of parts, or in the mode of distribution of differentiation
among them. Students, for example, of vertebrate morphology
debate whether the nth vertebra which carries the pelvic girdle
in Lizard A is individually homologous with the $n+x$th vertebra
which fulfils this function in Lizard B, or whether it is not more
truly homologous with the vertebra standing in the nth ordinal
position, though that vertebra in Lizard B is free.

In various and more complex aspects the same question is debated in regard to the cranial and spinal nerves, the branches of the aorta, the appendages of Arthropoda, and indeed in regard to all such series of differentiated parts in linear or successive repetition. Persons exercised with these problems should before making up their minds consider how similar questions would be answered in the case of any series of rhythmical repetitions formed by mechanical agencies. In the case of our illustration of the ripples in the sand, given the same forces acting on the same materials in the same area, the number of ripples produced will be the same, and the nth ripple counting from the end of the series will stand in the same place whenever the series is evoked. If any of the conditions be changed, the number and shapes can be changed too, and a fresh "distribution of differentiation" created. Stated in this form it is evident that the considerations which would guide the judgment in the case of the sand ripples are not essentially different from those which govern the problem of individual homology in its application to vertebrae, nerves, or digits.

The fact that the unit of repetition is also the unit of growth is the source of the obscurity which veils the process. When we compare the skeleton of a long-tailed monkey with that of a short-tailed or tailless ape we see at once how readily the additional series of caudal segments may be described as a consequence of the propagation of the "waves" of segmentation beyond the point where they die out in the shorter column, and we see that with an extension of the series of repetitions there is growth and extension of material.

The considerations which apply to this example will be found operating in many cases of the variation of terminal members of linear series. Some of these series, like the teeth of the dog, end in a terminal member of a size greatly reduced below that of the next to it. Even when there is thus a definite specialisation of the last member of the series it not infrequently happens that the addition, by variation, of a member beyond the normal terminal, is accompanied by a very palpable increase in size of the member which stands numerically in the place of the normal terminal.[2] So also with variation in the number of ribs, when a lumbar vertebra varies homoeotically into the likeness of the last dorsal and bears a rib, the rib placed next in front of this, which in the normal trunk is the last, shows a definite increase in development.

The consequences of such homoeoses are sometimes very extensive, involving readjustments of differentiation affecting a long series of members, as may easily be seen by comparing the vertebral columns of several individual Sloths[3] (whether *Bradypus* or *Choloepus*) to take a specially striking example.

It may be urged that no feature as yet enables us to perceive wherein lies the primary distinction which determines such variation, whether it is due to a difference in the dividing forces or in the material to be divided. If for instance we were to imitate such a series of segments by pressing hanging drops of a viscous fluid out of a paint-tube by successive squeezes, the number of times the tube is contracted before it is empty will give the number of the segments, but their size may depend either on the force of the contractions or on the capacity of the tube, or on various other factors. Nevertheless in the case of the variation of terminal members, whatever be the nature of the rhythmical impulse which produces the series of organs, the elevation of the normally terminal member in correspondence with the addition of another is what we should expect.

If the organism acquired its full size first and the delimitation of the parts took place afterwards, there might be some hope that the resemblance between living patterns and those mechanically caused by wave-motion might be shown to be a consequence of some real similarity of causation, but in view of the part played by growth, appeal to these mechanical phenomena cannot be declared to have more than illustrative value. Similarly in as much as living patterns appear, and almost certainly do in reality come into existence by a rhythmical process, comparisons of these patterns with those developed in crystalline structures, and in the various fields of force are, as it seems to me, inadmissible, or at least inappropriate.

[2] *Materials for the Study of Variation*, No. 249, p. 217; and p. 272.
[3] *Materials*, p. 118.

However their intermittence be determined, the rhythms of division must be looked upon as the immediate source of those geometrically ordered repetitions universally characteristic of organic life. In the same category we may thus group the segmentation of the Vertebrates and of the Arthropods, the concentric growth of the Lamellibranch shells or of Fishes' scales, the ripples on the horns of a goat, or the skeletons of the Foraminifera or of the Heliozoa. In the case of plant-structures Church[4] has admirably shown, with an abundance of detail, how on analysis the definiteness of phyllotaxis is an expression of such rhythm in the division of the apical tissues, and how the spirals and "orthostichies" displayed in the grown plant are its ultimate consequences. The problem thus narrows itself down to the question of the mode whereby these rhythms are determined.

It is natural that we should incline to refer them to a chemical source. If we think of the illustration just given, of the segmentation of a viscous fluid into drops by successive contractions of a soft-walled tube we can, I think, conceive of such rhythmic contractions as due to summations of chemical stimuli, somewhat as are the beats of the heart. But when we recognize the vast diversity of materials the distribution of which is determined by an ostensibly similar rhythmic process it seems hopeless to look forward to a directly chemical solution. That the chemical degradation of protoplasm or of materials which it contains is the source of the energy used in the divisions cannot be in dispute, but that these divisions can be themselves the manifestations of chemical action seems in the highest degree improbable.

We may therefore insist with some confidence on the distinction between the Meristic and the substantive constitution of organisms, between, that is to say, the system according to which the materials are divided and the essential composition of the materials, conscious of the fact that the energy of division is supplied from the materials, and that in the ontogeny the manner in which the divisions are effected must depend secondarily on the nature of the substances to be divided. The me-

[4] Church, A. H., *On the Relation of Phyllotaxis to Mechanical Laws*, London, 1904.

chanical processes of division remain a distinguishable group of phenomena, and variations in the substances to be distributed in division may be independent of variations in the system by which the distribution is effected.

Modern genetic analysis supplies many remarkable examples of this distinction. When formerly we compared the leaves of a normal palmatifid Chinese Primula with the pinnatifid leaves[5] of its fern-leaved variety we were quite unable to say whether the difference between the two types of leaf was due to a difference in the material cut up in the process of division or to a difference in that process itself. Knowledge that the distinction is determined by a single segregable factor tends to prove that the critical difference is one of substance. So also in the Silky fowl we know that the condition of its feathers is due to the absence of some one factor present in the normal form. We may conceive such differences as due to change of form in the successive "waves" of division, but we cannot yet imagine segregation otherwise than as acting by the removal or retention of a material element. Future observation by some novel method may suggest some other possibility, but such cases bring before us very clearly the difficulties by which the problem is beset.

In another region of observation phenomena occur which as it seems to me put it beyond question that the meristic forces are essentially independent of the materials upon which they act, save, in the remoter sense, in so far as these materials are the sources of energy. The physiology of those regenerations and repetitions which follow upon mutilation supplies a group of facts which both stimulate and limit speculation. No satisfactory interpretations of these extraordinary occurrences has ever been found, but we already know enough to feel sure that in them we are witnessing indications which should lead to the discovery of the true mechanics of repetition and pattern. The consequences of mutilation in causing new growth or perhaps more strictly in enabling new growth to take place, are such that they cannot be interpreted as responses to chemical stimuli in

[5] It is a question whether the dominance of the palmatifid leaf over the pinnatifid is not really an example of the dominance of a lower number of segmentations

FIG. 10. The palm- and fern type of leaf in *Primula Sinensis*. The palm is dominant and the fern is recessive.

any sense which the word chemical at present connotes. Powers are released by mutilation of which in the normal conditions of life no sign can be detected. All who have tried to analyse the phenomena of regeneration are compelled to have recourse to the metaphor of equilibrium, speaking of the normal body as in a state of strain or tension (Morgan) which when disturbed by mutilation results in new division and growth. The forces of division are inacessible to ordinary means of stimulation. Applications, for example, of heat or of electricty excite no responses of a positive kind unless the stimuli are so violent as to bring about actual destruction.[6] These agents do not, to use a loose expression, come into touch with the meristic forces. Changes in the chemical environment of cells may, as in the experiments of Loeb and of Stockard produce definite effects, but the facts suggest that these effects are due rather to alterations in the living material than to influence exerted directly on the forces of division themselves.

By destruction of tissue however the forces both of growth and of division also may often be called into action with a resulting regeneration. Interruption of the solid connexion between the parts may produce the same effects, as for example when the new heads or tails grow on the divided edges of Planarians (Morgan), or when from each half embryo partially separated from its normally corresponding half, a new half is formed with a twin monster as the result.

Often classed with regenerations but in reality quite distinct from them are those special and most interesting examples where the growth of a *paired* structure is excited by a simple

over a higher. From the uncertainty whether two given leaves of two separate plants are actually comparable one cannot institute quite satisfactory numerical comparisons, but I think the view that the "Fern" leaf has more lobes than an otherwise similar "Palm" leaf may be fairly maintained. If this be admitted, the "Palm" leaf represents the dominant low number and its round shape is a consequence of the greater powers of growth which are so often possessed by the members of a shorter series.

[6] It is perhaps of importance to remember that in certain species of bacteria (e. g. *Bacillus Anthracis*) division may cease where the organism is cultivated under certain artificial conditions though growth continues. In this way very long unsegmented threads are produced.

wound.   Some of the best known of these instances are presented
by the paired extra appendages of Insects and Crustacea.   Some
years ago I made an examination of all the examples of such
monstrosities to which access was to be obtained, and it was with
no ordinary feeling of excitement that I found that these supernumerary structures were commonly disposed on a recognizable
geometrical plan, having definite spatial relations both to each
other and to the normal limb from which they grew.   The more
recent researches of Tornier[7] and especially his experiments on
the Frog have shown that a cut into the posterior limb-bud
induces the outgrowth of such a *pair* of limbs at the wounded
place.   Few observations can compare with this in novelty or
significance; and though we cannot yet interpret these phenomena
or place them in their proper relations with normal occurrences,
we feel convinced that here is an observation which is no mere
isolated curiosity but a discovery destined to throw a new light
on biological mechanics.   The supernumerary legs of the Frog
are evidently grouped in a system of symmetry similar to that
which those of the Arthropods exhibit, and though in Arthropods
paired repetitions have not been actually produced by injury
under experimental conditions we need now have no hesitation
in referring them to these causes as Przibram has done.

At this point some of the special features of the supernumerary appendages become important.   First they may arise
at any point on the normal limb, being found in all situations
from the base to the apex.   Nor are they limited as to the surface
from which they spring, arising sometimes from the dorsal,
anterior, ventral, or posterior surfaces, or at points intermediate
between these principal surfaces.

With rare and dubious exceptions, the parts which are contained in these extra appendages are only those which lie *peripheral to their point of origin*.   Thus when the point of origin is
in the apical joint of the tarsus, the extra growth if completely
developed consists of a double tarsal apex bearing two pairs
of claws.   If they arise from the tibia, two complete tarsi are

    [7] *Arch. f. Entwm.*, XX, 1905, p. 76; *Sitzungsb. d. Ges. Naturf.*, Berlin, 1907,
p. 41, etc.

added. If they spring from the actual base of the appendage then two complete appendages may be developed in addition to the normal one. We must therefore conclude that in any point on a normal appendage the power exists which, if released, may produce a bud containing in it a paired set of the parts peripheral to this point.

Next the geometrical relations of the halves of the supernumerary pair are determined by the position in which they stand

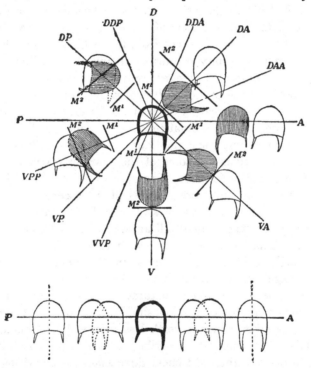

FIG. II. Diagrams of the geometrical relations which are generally exhibited by extra pairs of appendages in Arthropoda. The sections are supposed to be those of the apex of a tibia in a beetle. *A*, anterior, *P*, posterior, *D*, dorsal, *V*, ventral. M^1, M^2 are the imaginary planes of reflexion. The shaded figure is in each case a limb formed like that of the other side of the body, and the outer unshaded figures are shaped like the normal for the side on which the appendages are. On the several radii are shown the extra pairs in their several possible relations to the normal from which they arise. The normal is drawn in thick lines in the center.

in regard to the original appendage. These relations are best
explained by the diagram (Fig. 11), from which it will be seen
that the two supernumerary appendages stand as images of each
other; and, of them, that which is adjacent to the normal appendage forms an image of it. Thus if the supernumerary pair
arise from a point on the dorsal surface of the normal appendage,
the two *ventral* surfaces of the extra pair will face each other.
If they arise on the anterior surface of the normal appendage,
their morphologically posterior surfaces will be adjacent, and so on.

These facts give us a view of the relations of the two halves
of a dividing bud very different from that which is to be derived
from the exclusive study of normal structures. Ordinary morphological conceptions no longer apply. The distribution of the
parts shows that the bud or rudiment which becomes the supernumerary pair may break or open out in various ways according
to its relations to the normal limb. Its planes of division are
decided by its geometrical relations to the normal body.

Especially curious are some of the cases in which the extra
pair are imperfectly formed. The appearance produced is then
that of two limbs in various stages of coalescence, though in
reality of course they are stages of imperfect separation. The
plane of "coalescence" may fall anywhere, and the two appendages may thus be compounded with each other much as an
object partially immersed in mercury "compounds" with its
optical image reflected from the surface.

Supernumerary paired structures are not usually, if ever,
formed when an appendage is simply amputated. Cases occasionally are seen which nevertheless seem to be of this nature.
Borradaile,[8] for example, described a crab (*Cancer pagurus*)
having in place of the right chela three *small* chelae arising from
a common base, where the appearances suggested that the three
reduced limbs replaced a single normal limb. From the details reported however it seems still possible that one of the chelae
(that lettered F. 1 in Borradaile's figure) may be the normal
one, and the other two an extra pair. The chela which I suspect
to be the normal is in several respects deformed as well as being

[8] Borradaile, L. A., *Jour. Marine Zool.*, 1897, No. 8.

reduced in size, and this deformity may perhaps have ensued as a consequence of the same wound which excited the growth of the extra pair.   Its reduced size may be due to the same injury, which may quite well have checked its growth to full proportions.

Admitting doubt in these ambiguous cases it seems to be a general rule that for the production of the extra pair the normal limb should persist in connexion with the body.   Moreover it is practically certain that in no case can a *single*, viz. an unpaired, duplicate of the normal appendage grow from it.   Many examples have been described as of this nature, but all of them may be with confidence regarded as instances of a supernumerary pair in which only the two morphologically anterior or the two morphologically posterior surfaces are developed.   We have thus the paradox that a limb of one side of the body, say the right, has in it the power to form a pair of limbs, right and left, as an outgrowth of itself, but cannot form a second left limb alone.

A very interesting question arises whether it is strictly correct to describe the extra pair as a right and a left, or whether they are not rather two lefts or two rights of which one is reversed. This question did not occur to me when in former years I studied these subjects.   It was suggested to me by Dr. Przibram. The answer might have an important bearing on biological mechanics, but I know no evidence from which the point can be determined with certainty.   In order to decide this question it would be necessary to have cases in which the paired repetition affected a limb markedly differentiated on the two sides of the body, and of course the development of the extra parts in order to be decisive must be fairly complete.   One example only is known to me which at all satisfies these requirements, that of the lobster's chela figured (after Van Beneden) in *Materials for the Study of Variation*, p. 531, Fig. 184, III.

Here the drawing distinctly suggests that one of the extra dactylopodites, namely that lettered R, is differentiated as a left and not merely a reversed right.   For the teeth on this dactylopodite are those of a cutting claw, not of a crushing claw, whereas the dactylopodites R′ and L′ bear crushing teeth.   The figure makes it fairly certain also that the limb affected was a crushing claw. Accepting this interpretation, we reach the remarkable conclusion that the bud of new growth consisted of halves differentiated into cutter and crusher as the normal claws are, and that the extra crusher is geometrically a left but physiologically a right. Though shaped as a left in respect of the direction in which it points, the extra crusher is really an optically reversed right, while the dactylopodite R, which is placed pointing like a right, is really a reversed left (Fig. 12).

If these indications are reliable [9] and are established by further observation we shall be led to the conclusion that the bud which

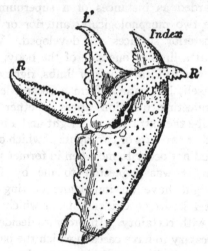

FIG. 12. Right claw of lobster bearing a pair of extra dactylopodites (after van Beneden). The fine toothing on R suggests that this is part of a cutting claw, though the limb bearing it is a crusher.

becomes an extra pair of limbs does not merely contain the parts proper to the side on which it grows, but is comparable with the original zygotic cell, and consists not simply of two halves, but of two halves differentiated as a right and a left like the two halves of the normal body.

Phenomena of this kind, evoked by mutilation or injury, together with the cognate observations on regeneration throw

[9] Dr. Przibram, I should mention, concludes that on the whole the facts are against this interpretation, but as more evidence is certainly required, I call attention to the possibility.

very curious lights on the nature of living things. To an understanding of the nature of the mechanics of living matter and its relation to matter at large they offer the most hopeful line of approach. I allude especially to the examples in which it has been established that the part which is produced after mutilation is a structure different from that which was removed. The term "regeneration" was introduced before such phenomena were discovered, and though every one recognizes its inapplicability to these remarkable cases, the word still misleads us by presenting a wrong picture to the mind. The expression "heteromorphosis" (Loeb) has been appropriately applied to various phenomena of this kind, and Morgan has given the name "morphallaxis" to another group of cases in which the renewal occurs by the transformation of a previously existing part.[10] But we must continually remember that all these occurrences which we know only as abnormalities and curiosities must in reality be exemplifications of the normal mechanics of division and growth. The conditions needed to call them forth are abnormal, but the responses which the system makes are evidences of its normal constitution. When therefore, for example, the posterior end of a worm produces a reversed tail from its cut end we have a proof that there must be in the normal body forces ready to cause this outgrowth. The new structure is not an ill-shaped head-end, for, as Morgan shows, the nephridial ducts have their funnels perforating the segments in a reversed direction. The "tension" of growth is actually reversed.[11] So also when in a Planarian amputation of the body immediately behind the head leads to the formation of a new reversed head at the back of the normal head, while amputation further back leads to the regeneration of a new tail, these responses give indications of forces normally present in the body of the Planarian. Such facts open up a great field of speculation and research. Especially important it would be to determine where the critical region may be at which the one response is replaced by the

[10] Morgan, T. H., *Regeneration*, 1901.

[11] It would be interesting to know whether growth continues at the original posterior end after the new "posterior" end has been formed in front.

other.   I suppose it is even possible that there is some neutral zone in which neither kind of response is made.

Physical parallels to the phenomena of regeneration are not easy to find and we still cannot penetrate beyond the empirical facts.   Przibram has laid stress on the general resemblance between the new growth of an amputated part in an animal and the way in which a broken crystal repairs itself when placed in the mother-solution.   That the two processes have interesting points of likeness cannot be denied.   It must however never be forgotten that there is one feature strongly distinguishing the two; for I believe it is universally recognized by physicists that all the phenomena of geometrical regularity which crystals display are ultimately dependent on the forms of the particles of the crystalline body.   This cannot in any sense be supposed to hold in regard to protoplasm or its constituents.   The definiteness of crystals is also an unlikely guide for the reason that it is absolute and perfect, or in other words because this kind of regularity cannot be disturbed at all without a change so great that the substance itself is altered; whereas we know that the forms of living things are capable of such changes, great and small, that we must regard perfection of form, whether manifested in symmetry or in number, as an ideal which will only be produced in the absence of disturbance.   The symmetry of the living things is like the symmetry of the concentric waves in a pool caused by a splash.   Perfect circles are made only in the imaginary case of mathematical uniformity, but the system maintains an approximate symmetry though liable to manifold deformation.

Since the geometrical order of the living body cannot be a direct function of the materials it must be referred to some more proximate control.   In renewing a part the body must possess the power of seizing particles of many dissimilar kinds, and whirl them into their several and proper places.   The action in renewal, like that of original growth, may be compared—very crudely—with the action of a separator which simultaneously distributes a variety of heterogeneous materials in an orderly fashion; but in the living body the thing distributed must rather be the *appetency* for special materials, not the materials themselves.

If the analogy of crystals be set aside and we seek for other parallels to regeneration there are none very obvious. I have sometimes wondered whether it might not be possible to institute a fruitful comparison between the renewal of parts and the reformation of waves of certain classes after obliteration. In several respects, as I have already said, some curious resemblances with the repetitions formed by wave-motion are to be traced in our organic phenomena, and though admitting that I cannot develop these comparisons, I think nevertheless they may be worth bearing in mind. When, after obliteration, an eddy in a stream, or a ripple-mark (a more complex case of eddy-formation) in blown sand is re-formed, we have an example in which pattern is reconstituted and growth takes place not by virtue of the composition of the materials—in this case the water or the sand— but by the way in which they are acted upon by extraneous forces.

A feature in the actual mode by which ripple-marks are reconstituted may not be without interest in connexion with our phenomena of regeneration. When, for example, the wind is blowing steadily over a surface of fine, dry sand, the familiar ripple-marks are formed by a heaping of the sand in lines transverse to the direction of the wind. The heaping is due to the formation of eddies corresponding with positions of instability. When the wind is steady and the sand homogeneous, the distances between the ripples, or wave-lengths, are sensibly equal. If while the wind continues to blow, the ripples are obliterated with a soft brush they will quickly be re-formed over the whole area, but I have noticed that at first their wave-length is approximately half that of the ripples in the undisturbed parts of the system.[12] The normal wave-length is restored by the gradual accentuation of alternate ripples. Of course the sand-ripples are in reality slowly travelling forward in the direction towards which the wind is blowing, and for this our living segmentations afford no obvious parallel, but the appearances in the area of

[12] In the actual case observed, the ripples unsmoothed had a wave-length of about 2½ inches; and when the new ones were first formed, there were about 30 ridges in the length originally traversed by 15 or 16.

reformation, and especially the forking of the old ridges where they join the new ones, are curiously reminiscent of the irregularities of segmentation seen in regenerated structures. The value of the considerations adduced in the chapter is, I admit, very small. The utmost that can be claimed for them is that mechanical segmentations, like those seen in ripple-mark, or in Leduc's osmotic growths, show how by the action of a continuous force in one direction, repeated and serially homologous divisions can be produced having features of similarity common to those repetitions by which organic forms and patterns are characterised. The analogy supplies a vicarious picture of the phenomena which in default of one more true may in a slight degree assist our thoughts. It suggests that the rhythms of segmentation may be the consequence of a single force definite in direction and continuously acting during the time of growth. The polarity of the organism would thus be the expression of the fact that this meristic force is definitely directed after it has once been excited, and the reversal seen in some products of regeneration suggest further that it is capable of being reflected. This polarity cannot be a property of the material, as such, but is determined by a force acting on that material, just as the polarity of a magnet is not determined by the arrangement of its particles, but by the direction in which the current flows.

To some it may appear that even to embark on such discussions as this is to enter into a perilous flirtation with vitalistic theories. How, they may ask, can any force competent to produce chemical and geometrical differentiation in the body be distinguished from the "Entelechy" of Driesch? Let me admit that in this reflexion there is one element of truth. If those who proclaim a vitalistic faith intend thereby to affirm that in the processes by which growth and division are effected in the body, a part is played by an orderly force which we cannot *now* translate into terms of any known mechanics, what observant man is not a vitalist? Driesch's first volume, putting as it does into intelligible language that positive deduction from the facts—especially of regeneration—should carry a vivid realisation of this truth to any mind. If after their existence is realised, it is desired that these unknown forces of order should have a name, and the word entelechy is proposed, the only objection I have to make is that the adoption of a term from Aristotelian philosophy carries a plain hint that we propose to relegate the future study of the problem to metaphysic.

From this implication the vitalist does not shrink. But I cannot find in the facts yet known to us any justification of so hopeless a course. It was but yesterday that the study of *Entwicklungsmechanik* was begun, and if in our slight survey we have not yet seen how the living machine is to be expressed in terms of natural knowledge that is poor cause for despair. Driesch sums up his argument thus:[13]

"It seems to me that there is only one conclusion possible. If we are going to explain what happens in our harmonious-equipotential sytems by the aid of causality based upon the constellation of single chemical factors and events, there *must* be some such thing as a machine. Now the assumption of the existence of a machine proves to be absolutely absurd in the light of the experimental facts. *Therefore there can be neither any sort of a machine nor any sort of causality based upon constellation underlying the differentiation of harmonious-equipotential systems.*

"For a machine, typical with regard to the three chief dimensions of space, cannot remain itself if you remove parts of it or if you rearrange its parts at will."

To the last clause a note is added as follows:

"The pressure experiments and the dislocation experiments come into account here; for the sake of simplicity they have not been alluded to in the main line of our argument."

I doubt whether any man has sufficient knowledge of all possible machines to give reality to this statement. In spite also of the astonishing results of experiments in dislocation, doubt may further be expressed as to whether they have been tried in such variety or on such a scale as to justify the suggestion that the living organism remains itself if its parts are rearranged at

[13] *The Science and Philosophy of the Organism;* Gifford Lectures, 1907. London, 1908, p. 141.

7

will. All we know is that it can "remain itself" when much is removed, and when much rearrangement has been affected, which is a different thing altogether.

I scarcely like to venture into a region of which my ignorance is so profound, but remembering the powers of eddies to re-form after partial obliteration or disturbance, I almost wonder whether they are not essentially machines which remain themselves when parts of them are removed.

Real progress in this most obscure province is not likely to be made till it attracts the attention of physicists; and though they for long may have to forego the application of exact quantitative methods, I confidently anticipate that careful comparison between the phenomena of repetition formed in living organisms and the various kinds of segmentation produced by mechanical agencies would be productive of illuminating discoveries.

CHAPTER IV

THE CLASSIFICATION OF VARIATION AND THE NATURE OF SUBSTANTIVE FACTORS

We have now seen that among the normal physiological processes the phenomena of division form a recognisable, and in all likelihood a naturally distinct group. Variations in these respects may thus be regarded as constituting a special class among variations in general.

The substantive variations have only one property in common—the negative one that they are not Meristic. The work of classifying them and distinguishing them according to their several types demands a knowledge of the chemistry of life far higher than that to which science has yet attained. In reference to some of the simplest variations Garrod has introduced the appropriate term "Chemical sports." The condition in man known as Alkaptonuria in which the urine is red is due especially to the absence of the enzyme which decomposes the excretory substance, alkapton. The "chemical sport" here consists in the inability to break up the benzene ring. The chemical feature which distinguishes and is the proximate cause of several colour-varieties can now in a few cases be declared. The work of Miss Wheldale has shown that colour-varieties may be produced by the absence of the chromogen compound the oxidation of which gives rise to sap-colours, by differences in the completeness of this process of oxidation, and by a process of reduction supervening on or perhaps suppressing the oxidation. Some of these processes moreover may be brought about by the combined action of two bodies, the one an enzyme, for example an oxygenase, and the other a substance regarded as a peroxide, contributing the oxygen necessary for the oxidation to take place. Variation in colour may thus be brought about by the addition or omission of any one of the bodies concerned in the action.

Similar variations, or rather similar series of variations will undoubtedly hereafter be identified in reference to all the various kinds of chemical processes upon which the structure and functions of living things depend. The identification of these processes and of the bodies concerned in them will lead to a real classification of Substantive Variations.

To forecast the lines on which such classification will proceed is to look too far ahead. We may nevertheless anticipate with some confidence that future analysis will recognise among the contributing elements, some which are intrinsic and inalienable, and others which are extrinsic and superadded.

We already know that there may be such interdependence among the substantive characters that to disentangle them will be a work of extreme difficulty. The mere fact that in our estimation characters belong to distinct physiological systems is no proof of their actual independence. In illustration may be mentioned the sap-colour in Stocks and the development of hoariness on the leaves and stems, which Miss Saunders's experiments have shown to be intimately connected, so that in certain varieties no hoariness is produced unless the elements for sap-colour are already present in the individual plant.

The first step in the classification of substantive variations is therefore to determine which are due to the addition of new elements or factors, and which are produced by the omission of old ones. *A priori* there is no valid criterion by which this can be known, and actual experiments in analytical breeding can alone provide the knowledge required. Some very curious results have by this method been obtained, which throw an altogether unexpected light on these problems. For example, in order that the remarkable development of mesoblastic black pigment characteristic of the Silky Fowl should be developed, it is practically certain that two distinct variations from such a type as *Gallus bankiva* must have occurred. I assume, as is reasonable, that *G. bankiva* has genetic properties similar to those of the Brown Leghorn breed which has been used in the experiments which Mr. Punnett and I have conducted. *Gallus bankiva* was not available but the Brown Leghorn agrees with it very closely in colouration, and probably in the general physiology of its pigmentation.

Setting aside the various structural differences between the two breeds, the Silky is immediately distinguished from the Leghorn by the fact that the skin of the whole body including that of the face and comb appears to be of a deep purplish colour. The face and comb of the Leghorn are red and the skin of the body is whitish yellow. On examination it is found that the purple colour of the Silky is in reality due to the distribution of a deep black pigment in the mesoblastic membranes throughout the body. The somatopleura, the pleura, *pia mater*, the dermis, and in most organs the connective tissue and the sheaths of the blood-vessels, are thus impregnated with black. No such pigmentation exists in the Leghorn. As the result of an elaborate series of experimental matings we have proved that the distinction between the Leghorn and the Silky consists primarily in the fact that the Silky possesses a pigment-producing factor, P, which is not present in the Leghorn.

This variation must undoubtedly have been one of *addition*. But besides this there is another difference of an altogether dissimilar nature; for the Brown Leghorn possesses a factor which has the power of partially or completely restricting the operation of the pigment-producing factor, P. Moreover in respect of this pigment-restricting factor which we may call D, the sexes of the Brown Leghorn differ, for the male is homozygous or DD, but the female is heterozygous, Dd. Thus in order that the black-skinned breed could be evolved from such a type as a Brown Leghorn it must be necessary *both* that P should be added *and* that D should drop out. We have not the faintest conception of the process by which either of these events have come to pass, but there is no reasonable doubt that in the evolution of the Silky fowl they did actually happen.

We may anticipate that numerous interdependences of this kind will be discovered.

Before any indisputable progress can be made with the problem of evolution it is necessary that we should acquire some real knowledge of the genesis of that class of phenomena which formed the subject of the last chapter. So long as the process of division remains entirely mysterious we can form no conception even of the haziest sort as to the nature of living organisms, or of the proximate causes which determine their forms, still less can we attempt any answer to those remoter questions of origin and destiny which form the subject of the philosopher's contemplation. It is in no spirit of dogmatism that I have ventured to indicate the direction in which I look for a solution, though I have none to offer. It may well be that before any solution is attained, our knowledge of the nature of unorganised matter must first be increased. For a long time yet we may have to halt, but we none the less do well to prepare ourselves to utilise any means of advance that may be offered, by carefully reconnoitering the ground we have to traverse. The real difficulty which blocks our progress is ignorance of the nature of division, or to use the more general term, of repetition.

Let us turn to the more familiar problem of the causes of variation. Now since variation consists as much in meristic change as in alteration in substance or material, there is one great range of problems of causation from which we are as yet entirely cut off. We know nothing of the causation of division, and we have scarcely an observation, experiment or surmise touching the causes by which the meristic processes may be altered.

Of the way in which variations in the substantive composition of organisms are caused we have almost as little real evidence, but we are beginning to know in what such variations must consist. These changes must occur either by the addition or loss of factors.

We must not lose sight of the fact that though the factors operate by the production of enzymes, of bodies on which these enzymes can act, and of intermediary substances necessary to complete the enzyme-action, yet these bodies themselves can scarcely be themselves genetic factors, but consequences of their existence. What then are the factors themselves? Whence do they come? How do they become integral parts of the organism? Whence, for example, came the power which is present in a White Leghorn of destroying—probably reducing—the pigment in its feathers? That power is now a definite possession of the breed, present in all its germ-cells, male and female, taking part in their symmetrical divisions, and passed on equally to all as much as is the protoplasm or any other attribute of the breed. From the body of the bird the critical and efficient substance could in all likelihood be isolated by suitable means, just as the glycogen of the liver can be. But even when this extraction has been accomplished and the reducing body isolated, we shall know no more than we did before respecting the mode by which the power to produce it was conferred on the fowl, any more than we know how the walls of its blood-vessels acquired the power to form a fibrin-ferment.

It is when the scope of such considerations as this are fully grasped that we realise the fatuousness of the conventional treatment which the problem of the causes of variation commonly receives. Environmental change, chemical injury, differences in food supply, in temperature, in moisture, or the like have been proposed as "causes." Admitting as we must do, that changes may be produced—usually inhibitions of development—by subjecting living things to changes in these respects, how can we suppose it in the smallest degree likely that very precise, new, and adaptative powers can be conferred on the germs by such treatment? Reports of positive genetic consequences observed comparable with those I have mentioned, become from time to time current. We should I think regard them with the gravest doubt. Few, so far as I am aware, have ever been confirmed, though clear and repeated confirmation should be demanded before we suffer ourselves at all to build upon such evidence. In a subsequent chapter some of these cases will be considered in detail.

In no class of cases would the transmission of an acquired character superficially appear so probable as in those where power of resisting the attack of a pathogenic organism is acquired in the lifetime of the zygote. The possession of such a power is moreover a distinction comparable with those which differentiate varieties and species. It is due to the development in the blood of specific substances which pervade the whole fluid. This development is exactly one of those "appropriate responses to stimuli" which naturalists who incline to regard adaptation as a direct consequence of an environmental influence might most readily invoke as an illustration of their views. And yet all evidence is definitely unfavourable to the suggestion of an inheritance of the acquired power of resistance. Such change as can be perceived in the virulence of the attacks on successive generations may be most easily regarded as due to the extermination of the more susceptible strains, and perhaps in some measure to variation in the invading organisms themselves, an "acquired character" of quite different import.

The specific "anti-body" may have been produced in response to the stimulus of disease, but the power to produce it without this special stimulus is not included in the germ-cells any more than a pigment. All that they bear is the *power to produce* the anti-bodies when the stimulus is applied.

If we could conceive of an organism like one of those to which disease may be due becoming actually incorporated with the system of its host, so as to form a constituent of its germ-cells and to take part in the symmetry of their divisions, we should have something analogous to the case of a species which acquires a new factor and emits a dominant variety. When we see the phenomenon in this light we realise the obscurity of the problem. The appearance of recessive varieties is comparatively easy to understand. All that is implied is the omission of a constituent. How precisely the omission is effected we cannot suggest, but it is not very difficult to suppose that by some mechanical fault of cell-division a power may be lost. Such variation by unpacking, or analysis of a previously existing complex, though unaccountable, is not inconceivable. But whence come the new dominants? Whether we imagine that they are created by some rearrangement or other change internal to the organism, or whether we try to conceive them as due to the assumption of something from without we are confronted by equally hopeless difficulty.

The mystery of the origin of a dominant increases when it is realised that there is scarcely any recent and authentic account of such an event occurring under critical observation, which can be taken as a basis for discussion. The literature of horticulture for example abounds in cases alleged, but I do not think anyone can produce an illustration quite free from doubt. Such evidence is usually open to the suspicion that the plant was either introduced by some accident, or that it arose from a cross with a pre-existing dominant, or that it owed its origin to the meeting of complementary factors. In medical literature almost alone however, there are numerous records of the spontaneous origin of various abnormal conditions in man which habitually behave as dominants, and of the authenticity of some of these there can be no doubt.

When we know that such conditions as hereditary cataract or various deformities of the fingers behave as dominants, we recognize that those conditions must be due to the addition of some element to the constitution of the normal man. In the collections of pedigrees relating to such pathological dominants there are usually to be found alleged instances of the origin of the condition *de novo*. Not only do these records occur with such frequency that they cannot be readily set aside as errors, but from general considerations it must be obvious that as these malformations are not common to normal humanity they must at some moment of time have been introduced. The lay reader may not be so much impressed with the difficulty as we are. He is accustomed to regard the origin of *any* new character as equally mysterious, but when once dominants are distinguished from recessives the problem wears a new aspect. Thus the appearance of high artistic gifts, whether as an attribute of a race or as a sporadic event among the children of parents destitute of such faculties, is not very surprising, for we feel fairly sure that the faculty is a recessive, due to the loss of a controlling or inhibiting factor; but the *de novo* origin of brachydactylous fingers in a child of normal parents is of quite a different nature, and must indicate the action of some new specific cause.

Whether such evidence is applicable to the general problem of evolution may with some plausibility be questioned; but there is an obvious significance in the fact that it is among these pathological occurrences that we meet with phenomena most nearly resembling the spontaneous origin of dominant factors, and I cannot see such pedigrees as these without recalling Virchow's aphorism that every variation owes its origin to some pathological accident. In the evolution of domestic poultry, if *Gallus bankiva* be indeed the parent form of all our breeds, at least some half dozen new factors must have been added during the process. In *bankiva* there is, for example, no factor for rose comb, pea comb, barring on the feathers, or for the various dominant types of dark plumage. Whence came all these? It is, I think, by no means impossible that some other wild species now extinct did take part in the constitution of domestic poultry. It seems indeed to me improbable that the heavy breeds descend from *bankiva*. Both in regard to domestic races of fowls, pigeons, and some other forms, the belief in origin within the period of human civilization from one simple primitive wild type seems on a balance of probabilities insecurely founded, but allowing something for multiplicity of origin we still fall far short of the requisite total of factors. Elements exist in our domesticated breeds which we may feel with confidence have come in since their captivity began. Such elements in fowls are dominant whiteness, extra toe, feathered leg, frizzling, etc., so that even hypothetical extension of the range of origin is only a slight alleviation of the difficulty.

Somehow or other, therefore, we must recognize that dominant factors do arise. Whether they are created by internal change, or whether, as seems to me not wholly beyond possibility, they obtain entrance from without, there is no evidence to show. If they were proved to enter from without, like pathogenic organisms, we should have to account for the extraordinary fact that they are distributed with fair constancy to half the gametes of the heterozygote.

In proportion as the nature of dominants grows more clear so does it become increasingly difficult to make any plausible suggestion as to their possible derivation. On the other hand the origin of a recessive variety by the loss of a factor is a process so readily imagined that our wonder is rather that the phenomenon is not observed far more often. Some slip in the accurate working of the mechanical process of division, and a factor gets left out, the loss being attested by the appearance of a recessive variety in some subsequent generation.

Consistently with this presentation of the facts we find that, as in our domesticated animals and plants, a diversity of recessives may appear within a moderately short period, and that when variations come they often do not come alone. Witness the cultural history of the Sweet Pea, *Primula Sinensis*, *Primula obconica*, *Nemesia strumosa* and many such examples in which variation when it did come was abundant. The fact cannot be too often emphasized that in the vast proportion of these examples of substantive variation under domestication, as well as of substantive variation in the natural state, the change has come about by omission, not by addition. To take, for example, the case of the Potato, in which so many spontaneous bud-variations have been recorded, East after a careful study of the evidence has lately declared his belief that all are of this nature, and the opinion might be extended to many other groups of cases whether of bud or seminal variation. Morgan draws the same conclusion in reference to the many varieties he has studied in *Drosophila*.

In the Sweet Pea, a form which is beyond suspicion of having been crossed with anything else, and has certainly produced all the multitude of types which we now possess by variations from one wild species, there is only one character of the modern types which could, with any plausibility, be referred to a factor not originally forming part of the constituents of the wild species. This is the waved edge, so characteristic of the "Spencer" varieties; for the cross between a smooth-edged and a waved type gives an intermediate not unfrequently. Nevertheless there is practically no doubt that this is merely an imperfection in the dominance of the smooth edge, and we may feel sure that any plant homozygous for smooth edge would show no wave at all. Hence it is quite possible that even the appearance of the original waved type, Countess Spencer, was due to the loss of one of the factors for smooth edge at some time in the history of the Sweet Pea.

In the case of the Chinese Primrose (*Primula Sinensis*) one dominant factor has been introduced in modern times, probably within the last six years at most. This is the factor which causes suppression of the yellow eye, giving rise to the curious type known as "Queen Alexandra." Mr. R. P. Gregory's experiments proved that this was a very definite dominant, and the element responsible for this development is undoubtedly an addition to the original ingredient-properties, with which the species was endowed. Unfortunately, as happens in almost every case of the kind, the origin of this important novelty appears to be lost. Its behaviour, however, when crossed with various other types is that of a simple dominant giving an ordinary 3 : 1 ratio. There is therefore no real doubt that it came into existence by the definite addition of a new factor, for if it was simply a case of the appearance of a new character made by combination of two previously existing complementary factors we should expect that when Queen Alexandra was self-fertilised a 9 : 7 ratio would be a fairly common result, which is not in practice found.

In *Oenothera* Gates[1] has observed the appearance, in a large sowing of about 1,000 *Oenothera rubrinervis*, of a single individual having considerably more red pigment in the calyx than is usual in *rubrinervis*. The whole of the hypanthium in the flowers of this plant was red instead of green as in *rubrinervis*, and the whole of the sepals were red in the bud-stage, except for small green areas at the base. This type behaved as a dominant over *rubrinervis*, but so far a pure-breeding individual was not found. Admittedly the variation of this plant from the type of *rubrinervis* can be represented as one of degree, though there is a very sensible gap in the series between the new form which Gates names "*rubricalyx*" and the reddest *rubrinervis* seen in his cultures. It must certainly be recognised as a new dominant. Gates, rightly as I consider, regards the distinction between *rubrinervis* and *rubricalyx* as a quantitative one, and the same remark applies to certain other types differing in the amount of anthocyanin which they produce. I do not understand the argu-

[1] Gates, R. R., *Zts. f. Abstammungslehre*, 1911, IV, pp. 341 and 361.

ment which Gates introduces to the effect that the difference between such quantitative types cannot be represented in terms of presence and absence. We are quite accustomed to the fact that in the rabbit self-colour segregates from the Dutch-marked type. These two types differ in a manner which we may reasonably regard as quantitative. It is no doubt possible that the self-coloured type contains an ingredient which enables the colour to spread over the whole body, but it is, I think, perhaps more easy to regard the Dutch type as a form from which a part of the colour is absent. It may be spoken of in terms I have used, as a *subtraction-stage* in colour. Following a similar method we may regard *rubricalyx* as an addition-stage in colour-variation. The fact that crosses between *rubrinervis*, or *rubricalyx* and *Lamarckiana* give a mixture of types in F_1, does not I think show, as Gates declares, that there is any system here at work to which a factorial or Mendelian analysis does not apply; but that question may be more fitly discussed in connexion with the other problems raised by the behaviour of *Oenothera* species in their crosses.

I do, however, feel that, interesting as this case must be admitted to be, we cannot quite satisfactorily discuss it as an illustration of the *de novo* origin of a dominant factor. The difference between the novelty and the type is quantitative, and it is not unreasonable to think of such a difference being brought about by some "pathological accident" in a cell-division.

Recognition of the distinction between dominant and recessive characters has, it must be conceded, created a very serious obstacle in the way of any rational and concrete theory of evolution. While variations of all kinds could be regarded as manifestations of some mysterious instability of organisms this difficulty did not occur to the mind of evolutionists. To most of those who have taken part in genetic analysis it has become a permanent and continual obsession. With regard to the origin of recessive variations, there is, as we have seen, no special difficulty. They are negative and are due to absences, but as soon as it is understood that dominants are caused by an addition we are completely at a loss to account for their origin, for we cannot surmise any source from which they may have been derived. Just as when typhoid fever breaks out in his district the medical officer of health knows for certain that the bacillus of typhoid fever has by some means been brought into that district so do we know that when first dominant white fowls arose in the evolution of the domestic breeds, by some means the factor for dominant whiteness got into a bird, or into at least one of its germ-cells. Whence it came we cannot surmise.

Whether we look to the outer world or to some rearrangement within the organism itself, the prospect of finding a source of such new elements is equally hopeless.

Leaving this fundamental question aside as one which it is as yet quite unprofitable to discuss, we are on safe ground in foreseeing that the future classification of substantive variations, which genetic research must before long make possible, will be based on a reference to the modes of action of the several factors. Some will be seen to produce their effects by oxidation, some by reduction, some by generating substances of various types, sugars, enzymes, activators, and so forth. It may thus be anticipated that the relation of varieties to each other and to types from which they are derived will be expressible in terms of definite synthetical formulae. Clearly it will not for an indefinite time be possible to do this in practice for more than a few species and for characters especially amenable to experimental tests, but as soon as the applicability of such treatment is generally understood the influence on systematics must be immediate and profound, for the nature of the problem will at length be clear and, though the ideal may be unattainable, its significance cannot be gainsaid.

Note.—With hesitation I allow this chapter to appear in the form in which it was printed a year ago, but in passing it for the press after that interval I feel it necessary to call attention to a possible line of argument not hitherto introduced.

In all our discussions we have felt justified in declaring that the dominance of any character indicates that some factor is present which is responsible for the production of that character. Where there is no definite dominance and the heterozygote is of an intermediate nature we should be unable to declare on which side the factor concerned was present and from which side it was absent. The degree of dominance becomes thus the deciding criterion by which we distinguish the existence of factors. But it should be clearly realized that in any given case the argument can with perfect logic be inverted. We already recognize cases in which by the presence of an inhibiting factor a character may be suppressed and purely as a matter of symbolical expression we might apply the same conception of inhibition to any example of factorial influence whatever. For instance we say that in as much as two normal persons do not have brachydactylous children, there must be some factor in these abnormal persons which causes the modification. Our conclusion is based on the observed fact that the modification is a dominant. But it may be that normal persons are homozygous in respect of some factor N, which prevents the appearance of brachydactyly, and that in any one heterozygous, Nn, for this inhibiting factor, brachydactyly can appear. Similarly the round pea we say contains R, a factor which confers this property of roundness, without which its seeds would be wrinkled. But here we know that the wrinkled seed is in reality one having compound starch-grains, and that the heterozygote, though outwardly round enough, is intermediate in that starch-character. If we chose to say that the compoundness of the grains is due to a factor C and that two doses of it are needed to make the seed wrinkled, I know no evidence by which such a thesis could be actually refuted. That such reasoning is seemingly perverse must be conceded; but when we consider the extraordinary difficulties which beset any attempt to conceive the mode of origin of a new dominant factor, we are bound to remember that there is this other line of argument which avoids that difficulty altogether. In the case of the " Alexandra "-eye in *Primula*, or the red calyx in Gates's *Oenothera*, inverting the reasoning adopted in the text, we may see that only the *Primula* homozygous for the yellow eye can develop it and that two doses of the factor for the *rubrinervis* calyx are required to prevent that part of the plant from being red.

We may proceed further and extend this mode of reasoning to all cases of genetic variation, and thus conceive of all alike as due to loss of factors present in the original complex. Until we can recognize factors by means more direct than are provided by a perception of their effects, this doubt cannot be positively removed. For all practical purposes of symbolic expression we may still continue to use in our analyses the modes of representation hitherto adopted, but we must not, merely on the ground of its apparent perversity, refuse to admit that the line of argument here indicated may some day prove sound.

CHAPTER V

THE MUTATION THEORY

When with the thoughts suggested in the last chapter we contemplate the problem of Evolution at large the hope at the present time of constructing even a mental picture of that process grows weak almost to the point of vanishing. We are left wondering that so lately men in general, whether scientific or lay, were so easily satisfied. Our satisfaction, as we now see, was chiefly founded on ignorance.

Every specific evolutionary change must represent a definite event in the construction of the living complex. That event may be a disturbance in the meristic system, showing itself in a change in the frequency of the repetitions or in the distribution of differentiation among them, or again it may be a chemical change, adding or removing some factor from the sum total.

If an attempt be made to apply these conceptions to an actual series of allied species the complexity of the problem is such that the mind is appalled. Ideas which in the abstract are apprehended and accepted with facility fade away before the concrete case. It is easy to imagine how Man was evolved from an *Amoeba*, but we cannot form a plausible guess as to how *Veronica agrestis* and *Veronica polita* were evolved, either one from the other, or both from a common form. We have not even an inkling of the steps by which a Silver Wyandotte fowl descended from *Gallus Bankiva*, and we can scarcely even believe that it did. The Wyandotte has its enormous size, its rose comb, its silver lacing, its tame spirit, and its high egg production. The tameness and the high egg production are probably enough both recessives, and though we cannot guess how the corresponding dominant factors have got lost, it is not very difficult to imagine that they were lost somehow. But the rose comb and the silver colour are *dominants*. The heavy weight also appears in the crosses with Leghorns, but we need not at once conclude that it

8

depends on a simple dominant factor, because the big size of
the crosses may be a consequence of the cross and may depend on
other elements.

Now no wild fowl known to us has these qualities. May we
suppose that some extinct wild species had them? If so, may
we again make the same supposition in all similar cases? To do
so is little gain, for we are left with the further problem, whence
did those lost wild species acquire those dominants? Supposi-
tions of this kind help no more than did the once famous
conjecture as to the origin of living things—that perhaps they
came to earth on a meteorite. The unpacking of an original
complex, the loss of various elements, and the recombination of
pre-existing materials may all be invoked as sources of specific
diversity. Undoubtedly the range of possibilities thus opened
up is large. It will even cover an immense number of actual
examples which in practice pass as illustrations of specific distinction. The Indian Rock pigeon which has a blue rump
may quite reasonably be regarded as a geographically separated
recessive form of our own *Columba livia*, for as Staples-Browne
has shown the white rump of *livia* is due to a dominant factor.
The various degrees to which the leaves of Indian Cottons are
incised have, as Leake says, been freely used as a means of classification. The diversities thus caused are very remarkable,
and when taken together with diversities in habit, whether
sympodial or monopodial, the various combinations of points
of difference are sufficiently distinctive to justify any botanist
in making a considerable number of species by reference to them
alone. Nevertheless Leake's work goes far to prove that all of
these forms represent the re-combinations of a very small number
of factors. The classical example of *Primula Sinensis* and its
multiform races is in fact for a long way a true guide as to the
actual interrelations of the species which systematists have
made. That they did make them was due to no mistake in
judgment or in principle, but simply to the want of that extended knowledge of the physiological nature of the specific
cases which we now know to be a prime necessity.

But will such analysis cover all or even most of the ordinary

cases of specific diversity between near allies? Postponing the
problem of the interrelations of the larger divisions as altogether
beyond present comprehension, can we suppose, that in general,
closely allied species and varieties represent the various consequences of the presence or absence of allelomorphic factors
in their several combinations? The difficulty in making a
positive answer lies in the fact that in most of the examples in
which it has been possible to institute breeding experiments with
a view to testing the question, a greater or less sterility is encountered. Where, however, no such sterility is met with, as
for instance in the crosses made by E. Baur among the species
of *Antirrhinum* there is every reason to think that the whole
mass of differences can and will eventually be expressed in terms
of ordinary Mendelian factors. Baur has for example crossed
species so unlike as *Antirrhinum majus* and *molle*, forms differing
from each other in almost every feature of organisation.[1] The
F_2 generation from this cross presents an amazingly motley
array of types which might easily if met with in nature be described as many distinct species. Yet all are fertile and there
is not the slightest difficulty in believing that they can all be
reduced to terms of factorial analysis.

If allowance be made for the complicating effects of sterility,
is there anything which prevents us from supposing that such
good species as those of *Veronica* or of any other genus comprising
well-defined forms may not be similarly related? I do not know
any reason which can be pointed to as finally excluding such a
possibility. Nevertheless it has been urged with some plausibility that good species are distinguished by *groups* of differentiating characters, whereas if they were really related as the
terms of a Mendelian F_2 family are, we should expect to find
not groups of characters in association, but rather series of forms
corresponding to the presence and absence of the integral factors
composing the groups of characters. I am not well enough
versed in systematic work to be able to decide with confidence
how much weight should be attached to this consideration. Some

[1] See Lotsy and Baur, Rep. Genetics Conf., Paris, 1911, pp. 416–426. Compare Lecoq on *Mirabilis jalapa* × *longiflora*, Fécondation des Végétaux, 1862, p. 311.

weight it certainly has, but I cannot yet regard it as forming a fatal objection to the application of factorial conceptions on the grand scale. It may be recalled that we are no longer under any difficulty in supposing that differences of all classes may be caused by the presence or absence of factors. It seemed at first for example that such characters as those of leaf shape might be too subtle and complex to be reducible to a limited number of factors. But first the work of Gregory on *Primula Sinensis* showed that several very distinct types of leaves were related to each other in the simplest way. In that particular example, intermediates are so rare as to be negligible, but subsequently Shull dealing with such a complicated example as *Capsella*, and Leake in regard to Cottons, both forms in which intergrades occur in abundance, have shown that a simple factorial scheme is applicable. We need not therefore, to take an extreme case, doubt that if it were possible to examine the various forms of fruit seen in the Squashes by really comprehensive breeding tests, even this excessive polymorphism in respect of structural features would be similarly reducible to factorial order.

It must always be remembered also that in a vast number of cases, nearly allied forms which are distinct, occupy distinct ground. Moreover, by whatever of the many available mechanisms that end be attained, it is clear that nature very often does succeed in preventing intercrossing between distinct forms so far that the occurrence of that phenomenon is a rarity under natural conditions. The facts may, I think, fairly be summarized in the statement that species are on the whole distinct and not intergrading, and that the distinctions between them are usually such as might be caused by the presence, absence, or inter-combination of groups of Mendelian factors; but that they are so caused the evidence is not yet sufficient to prove in more than a very few instances.

The alternative, be it explicitly stated, is not to return to the view formerly so widely held, that the distinctions between species have arisen by the accumulation of minute or insensible differences. The further we proceed with our analyses the more inadequate and untenable does that conception of evolutionary change become. If the differences between species have not come about by the addition or loss of factors one at a time, then we must suppose that the changes have been effected by even larger steps, and variations including groups of characters, must be invoked.

That changes of this latter order are really those by which species arise, is the view with which de Vries has now made us familiar by his writings on the Mutation Theory. In so far as mutations may consist in meristic changes of many kinds and in the loss of factors it is unnecessary to repeat that we have abundant evidence of their frequent occurrence. That they may also more rarely occur by the addition of a factor we are, I think, compelled to believe, though as yet the evidence is almost entirely circumstantial rather than direct. The evidence for the occurrence of those mutations of higher order, by which new species characterized by several distinct features are created, is far less strong, and after the best study of the records which I have been able to make, I find myself unconvinced. The facts alleged appear capable of other interpretations.

The most famous and best studied examples are of course the forms of *Oenothera* raised by de Vries from *Oenothera Lamarckiana* in circumstances well known to all readers of genetic literature. Whatever be the true significance of these extraordinary "mutations" there can be no question about the great interest which attaches to them, and the historical importance which they will long preserve. Apart also from these considerations it is becoming more and more evident that in their peculiarities they provide illustrations of physiological phenomena of the highest consequence in the study of genetics at large.

De Vries found, as is well known, that *Oenothera Lamarckiana* gives off plants unlike itself. These mutational forms are of several distinct and recognizable types which recur, and several of them breed true from their first appearance. The obvious difficulty, which in my judgment should make us unwilling at present to accept these occurrences as proof of the genesis of new species by mutation, is that we have as yet no certainty that the appearance of the new forms is not an effect of the recombination of factors, such as is to be seen in so many generations of plants derived from a cross involving many genetic elements. The first question is what is *Oenothera Lamarckiana?* Is it itself a plant of hybrid origin? To this fundamental question no satisfactory answer has yet been given. All attempts to find it as a wild plant in America have failed. It existed in Europe in the latter half of the eighteenth century. Whence it came is still uncertain, but the view that it came into existence in Europe and perhaps in Paris, seems on the whole the most probable. The question has been debated by Macdougal, Gates, and Davis. From historical sources there is little expectation of further light. Those who favour the notion of a hybrid origin look on *Oenothera biennis* as one of the putative parents. It has been conjectured that a species called *grandiflora* lately re-discovered on the Alabama river was the other parent. Experiments have been instituted by Davis to discover whether *Lamarckiana* can be made artificially by crossing these two species. The results so far have shown that while plants approximating in various respects to *Lamarckiana* have thus been produced, none agree exactly with that form. Davis, to whom reference should be made for a full account of the present state of the enquiry, points out that there are many strains of *biennis* in existence and that it is by no means impossible that by using others of these strains a still closer approximation can be made. None of Davis's artificial productions as yet breed at all true, as *Lamarckiana* on the whole does. In such a case, however, where several characters are involved, this is perhaps hardly to be expected.

One feature of the *Oenotheras* is very curious. Not only *Lamarckiana*, but all the allied species so far as I am aware, have a considerable proportion of bad and shrivelled pollen grains. This is undoubtedly true of species living in the wild state as well as of those in cultivation. I have had opportunities of verifying this for myself in the United States. No one looking at the pollen of an *Oenothera* would doubt that it was taken from some hybrid plant exhibiting partial sterility. On the other hand, it is difficult to suppose that numbers, perhaps all, of the "species" of the genus are really hybrids, and many of them breed substantially true. I regard this constant presence of bad pollen grains as an indication that the genetic physiology of *Oenothera* is in some way abnormal, and as we shall presently see, there are several other signs which point in the same direction.

Discussion of the whole series of phenomena is rendered exceedingly difficult first, by reason of the actual nature of the material. The characteristics of many of the types which de Vries has named are evasive. A few of these types, for instance, *gigas*, *nanella*, *albida*, *brevistylis*, and perhaps a few more are evidently clear enough, but we have as yet no figures and descriptions precise enough to enable a reader to appreciate exactly the peculiarities of the vast number of forms which have now to be considered in any attempt to gain a comprehensive view of the whole mass of facts. It is also not in dispute that the forms are susceptible of great variations due simply to soil and cultural influences.

The fact that no Mendelian analysis has yet been found applicable to this group of *Oenotheras* as a whole is perhaps largely due to the fact that until recently such analysis has not been seriously attempted. Following the system which he had adopted before the rediscovery of Mendelism, or at all events, before the development of that method of analysis, de Vries has freely applied *names* to special combinations of characters and has scarcely ever instituted a factorial analysis. Before we can get much further this must be attempted. It may fail, but we must know exactly where and how this failure comes about. There are several indications that such a recognition of factorial characters, could be carried some way. For example, the height, the size of the flowers, the crinkling of the leaves, the brittleness of the stems, perhaps even the red stripes on stems and fruits, and many more, are all characters which may or may not depend on distinct factors, but if such characters are really transmitted in unresolved groups, the limitations of those groups should be carefully determined. The free use of names for the several forms, rather than for the characters, has greatly contributed to deepen the obscurity which veils the whole subject.

I do not mean to suggest that these *Oenotheras* follow a simple Mendelian system. All that we know of them goes to show that there are curious complications involved. One of these, probably the most important of all, has lately been recognized by de Vries himself, namely, that in certain types the characters borne by the female and the male germ-cells of the same plant are demonstrably different. There can be little doubt that further research will reveal cognate phenomena in many unsuspected places. The first example in which such a state of things was proved to exist is that of the Stocks investigated by Miss Saunders.[2] By a long course of analysis she succeeded in establishing in 1908 the fact that if a plant of *Matthiola* is of that eversporting kind which gives a large proportion of double-flowered plants among its offspring (produced by self-fertilisation), then the egg-cells of such a plant are mixed in type, but the pollen of the same plant is homogeneous. Some of the egg-cells have in them the two factors for singleness, but some of them are short of one or both of these factors. The pollen-grains, however, are all recessives, containing neither of these factors. The egg-cells, in other words, are mixed, "singles" and "doubles," while the pollen-grains are all "doubles." The same is true of the factor differentiating "white," or colourless plastids from cream-coloured plastids in *Matthiola*, the egg-cells being mixed "whites" and "creams," while the pollen-grains are all "creams," viz: recessives. Later in the same year (1908) de Vries[3] announced a remarkable case which will be discussed in detail subsequently. It relates to certain *Oenotheras* heterozygous for dwarfness, in which (p. 113) the ovules were mixed, talls and dwarfs, while the pollen is all dwarf.

Again in *Petunia* Miss Saunders's[4] work has shown that a somewhat similar state of things exists, but with this remarkable difference, that though the egg-cells are mixed, singles and doubles, the pollen-grains are all *singles*, viz: dominants. All the *Petunias* yet examined have been in this condition, including

[2] *Rep. Evol. Ctee. R. S.*, IV, 1908, p. 38.
[3] *Ber. Deut. Bot. Ges.*, 1908, XXVI, a, p. 672.
[4] *Jour. Genetics*, I, 1910, p. 57.

some which in botanic gardens pass for original species. Whether
actual wild plants from their native habitats are in the same
state, is not yet known, but it is by no means improbable. The
case may be compared with that of the moth *Abraxas grossulariata* studied by Doncaster and Raynor, in which the females
are all heterozygous, or we may almost say "hybrids" of *grossulariata* and the variety *lacticolor*. Similarly we may say that at
least garden Petunias are heterozygous in respect of singleness.
The proof of this is of course that when fertilised with the pollen
of doubles they throw a mixture of doubles and singles. The
statements which de Vries has published regarding the behaviour
of several of the *Oenotheras* go far to show that they must have
a somewhat similar organisation. On the present evidence it is
still quite impossible to construct a coherent scheme which will
represent all the phenomena in their interrelations, and among
the facts are several which, as will appear, seem mutually incompatible. The first indication that the *Oenotheras* may have
either mixed ovules or mixed pollen appears in the fact that
Lamarckiana and several of its "mutants" used as males, with
several other forms as females, give a mixed offspring. For
example, de Vries (1907) found that

> *biennis* ♀ × *Lamarckiana* ♂
> *biennis cruciata* ♀ × *Lamarckiana* ♂
> *muricata* ♀ × *Lamarckiana* ♂
> *biennis* ♀ × *rubrinervis* ♂
> *biennis cruciata* ♀ × *rubrinervis* ♂

all give a mixture of two distinct types which he names *laeta*
and *velutina*, consisting of about equal numbers of each. On
account of the fact that the two forms are produced in association
de Vries has called these forms "twin hybrids," a designation
which is not fortunate, seeing that it is impossible to imagine
that any kind of twinning is concerned in their production. The
distinction between these two seems to be considerable, *laeta*
having leaves broader, bright green in colour, and flat, with
pollen scanty, while *velutina* has leaves narrower, grayish green,
more hairy, and furrow-shaped, with pollen abundant.

We next meet the remarkable fact that these two forms,

laeta and *velutina* breed true to their respective types, and do not reproduce the parent-types among their offspring resulting from self-fertilisation. This statement must be qualified in two respects. When *muricata* ♂ is fertilised by *brevistylis* the forms *laeta* and *velutina* are produced, but each of them subsequently throws the short-styled form as a recessive (de Vries, 1907, p. 406). It may be remembered that de Vries's previous publications had already shown that the short style of *brevistylis*, one of the *Lamarckiana* "mutants," behaves as a recessive habitually (*Mutationstheorie*, II, p. 178, etc.).

Also when *nanella*, the dwarf "mutant" of *Lamarckiana* is used as male on *muricata* as female, *laeta* and *velutina* are produced, but one only of these, namely, *velutina*, subsequently throws dwarfs on self-fertilisation. The dwarfs thus thrown are said to form about 50 per cent. of the families in which they occur (de Vries, 1908, p. 668). The fact that the two forms, *laeta* and *velutina*, are produced by many matings in which *Lamarckiana* and its mutant *rubrinervis* are used as males is confirmed abundantly by Honing, who has carried out extensive researches on the subject. After carefully reading his paper, I have failed to understand the main purport of the argument respecting the "double nature" of *Lamarckiana* which he founds on these results, but I gather that in some way *laeta* is shown to partake especially of the nature of *Lamarckiana*, while *velutina* is a form of *rubrinervis*. The paper contains many records which will be of value in subsequent analysis of these forms.

Before considering the possible meaning of these facts we must have in our minds the next and most novel of the recent extensions of knowledge as to the genetic properties of the *Oenotheras*. In the previous statement we have been concerned with the results of using either *Lamarckiana* itself or one of its "mutants" *rubrinervis*, *brevistylis*, or *nanella* as male, on one of the species *biennis* or *muricata*. The new experiments relate to crosses between the two species *biennis* and *muricata* themselves.

De Vries found:

1. That the reciprocal hybrids from these two species differed,

biennis × *muricata* producing one type of F_1 and *muricata* × *biennis* producing another. Each F_1 resembled the father more than the mother.

2. That each of the hybrids so produced breeds true on self-fertilisation.

3. That if we speak of the hybrid from *biennis* × *muricata* as *BM* and of the reciprocal as *MB*, then

$$BM \times MB$$

gives exclusively offspring of *biennis* type but that

$$MB \times BM$$

gives exclusively offspring of *muricata* type. Evidently, apart from all controversy as to the significance of the "mutants" of *Lamarckiana*, we have here a series of observations of the first importance.

The fact that reciprocal crossings give constantly distinct results must be taken to indicate that the male and female sides of one, if not of both, of the parents are different in respect of characters which they bear. This is de Vries's view, and he concludes rightly, I think, that the evidence from all the experiments shows that both *biennis* and *muricata* are in this condition, having one set of characters represented in their pollen-grains and another in their ovules. The plants breed true, but their somatic structures are compounded of the two sets of elements which pass into them from their maternal and paternal sides respectively. This possibility that species may exist of which the males really belong to one form and the females to another, is one which it was evident from the first announcement of the discovery of Mendelian segregation might be found realised in nature.[5]

Oe. biennis and *muricata* were crossed reciprocally with each other and with a number of other species, and the behaviour of each, when used as mother, was consistently different from its behaviour when used as father. De Vries is evidently justified

[5] In Rep. 1 to Evol. Committee, 1902, p. 132, attention was called to this possibility, though of course at that date it was in sexual animals alone that it was supposed to exist. It had not occurred to me that even a hermaphrodite plant might be in this condition.

108 PROBLEMS OF GENETICS

by the results of this series of experiments in stating that the
"Bild," as he terms it, or composition of the male and female
sides of these two species, *biennis* and *muricata*, are distinct.
On the evidence before us it is not, however, possible to form a
perfectly clear idea of each, and until details are published, a
reader without personal knowledge of the material cannot do
more than follow the general course of the argument. For fuller
comprehension a proper analysis of the characters with a clear
statement of how they are distributed among the several types
and crosses is absolutely necessary. According to de Vries the
female of *biennis* possesses a group of characters which he defines
as "*conica*" in allusion to the shape of the flower-buds. Besides
the conical buds, this group of features includes imperfect
development of wood, rendering the plant very liable to attacks
of *Botrytis*, and comparatively narrow leaves.

The female of *muricata* carries a group of features which he
calls "*frigida*," and, though this is not quite explicitly stated in a
definition of that type, it is to be inferred[6] that its characteristics
are regarded as greater height, strong development of wood with
comparative resistance to *Botrytis*, and broad leaves.

The characters borne by the male parts of the two species
are in general those by which they are outwardly distinguished.
For example, the leaves of *Oe. biennis* are comparatively broad
and are bright green, while those of *muricata* are much narrower
and of a glaucous green, and I understand that de Vries regards
these properties as contributed by the male side in each case and
to be carried by the male cells of each species. The suggestion
as regards *biennis* and *muricata* comes near the conception often
expressed by naturalists in former times (*e. g.*, Linnaeus) and
not rarely entertained by breeders at the present day, that the
internal structure is contributed by the mother and the external
by the father.

On the other hand, the offspring of each species when used
as mother is regarded as possessing in the main the features of
the maternal "Bild," but the matter is naturally complicated
by the introduction of features from the father's side, and it is

_____
[6] From the description of the offspring of *muricata* used as mother.

here especially that the account provided is at present unsatisfactory and inconclusive. There seems, however, to be no serious
doubt that *biennis* and *muricata* each in their outward appearance
exhibit on the whole the features which their pollens respectively
carry, and that the features borne by their ovules are in many
respects distinct.

The *types* are thus "hybrids" which breed true. The results
of intercrossing them each way are again "hybrids" which breed
true. It will be remembered that on former occasions de Vries
has formulated a general rule that *species*-hybrids breed true,
but that the cross-breds raised by interbreeding *varieties* do not.
One of these very cases was quoted[7] as an illustration of this
principle, viz: *muricata* × *biennis*. The grounds for this general
statement have always appeared to me insufficient, and with the
further knowledge which the new evidence provides we are
encouraged to hope that when a proper factorial analysis of the
types is instituted we shall find that the phenomenon of a constant hybrid will be readily brought into line with the systems
of descent already worked out for such cases as that of the Stocks,
and others already mentioned.

In further discussion of these facts de Vries makes a suggestion
which seems to me improbable. Since the egg-cells of *muricata*,
for instance, bear a certain group of features which are missing
on the male side, and conversely the pollen bears features absent
from the female side, he is inclined to regard the *bad pollen grains*
as the bearers of the missing elements of the male side and to
infer that there must similarly be defective ovules representing
the missing elements of the female side. No consideration is
adduced in support of this view beyond the simple fact that the
characters borne by male and female are dissimilar, whereas
it would be more in accord with preconception if the same sets
of combinations were represented in each—as in a normal
Mendelian case. There is as yet no instance in which the absence
of any particular class of gametes has been shown with any
plausibility to be due to defective viability, though there are, of
course, cases in which certain classes of zygotes do not survive

[7] de Vries, *Species and Varieties*, 1905, p. 259.

owing to defective constitution (*e. g.*, the albinos of *Antirrhinum* studied by Baur, and the homozygous yellow mice). I am rather inclined to suppose that in these examples of hybrids breeding true we shall find a state of things comparable with that to which we formerly applied the terms "coupling" and "repulsion." In these cases certain of the possible combinations of factors occur in the gametic series with special frequency, being in excess, while the gametes representing other combinations are comparatively few. In a recent paper on these cases Professor Punnett and I have shown that these curious results vary according to the manner in which the factors are grouped in the parents. If A and B are two factors which exhibit these phenomena we find that the gametic series of the double heterozygote differs according as the combination is made by crossing $AB \times ab$, or by crossing $Ab \times aB$. In a normal Mendelian case the F_1 form, $AaBb$, produces gametes AB, Ab, aB, ab, in equal numbers; but in these peculiar cases those gametes which contain

	Gametic series				Number of gametes in series	Number of zygotes formed
	AB	Ab	aB	ab		
Partial repulsion from zygote of form $Ab \times aB$	1	$(n-1)$	$(n-1)$	1	$2n$	$4n^2$
	1	31	31	1	64	4096
	1	15	15	1	32	1024
	1	7	7	1	16	256
	1	3	3	1	8	64
	1	1	1	1	4	16
Partial coupling from zygote of form $AB \times ab$	3	1	1	3	8	64
	7	1	1	7	16	256
	15	1	1	15	32	1024
	31	1	1	31	64	4096
	63	1	1	63	128	16384
	$(n-1)$	1	1	$(n-1)$	$2n$	$4n^2$

	Nature of zygotic series			
	AB	Ab	aB	ab
Partial repulsion from zygote of form $Ab \times aB$	$2n^2+1$	n^2-1	n^2-1	1
	2049	1023	1023	1
	513	255	255	1
	129	63	63	1
	33	15	15	1
	9	3	3	1
Partial coupling from zygote of form $AB \times ab$	41	7	7	9
	177	15	15	49
	737	31	31	225
	3009	63	63	961
	12161	127	127	3969
	$3n^2-(2n-1)$	$2n-1$	$2n-1$	$n^2-(2n-1)$

the *parental combinations* are in excess. This excess almost
certainly follows the system indicated by the accompanying
table. In the general expressions n is half the number of gametes
required to express the whole system. Now if we imagine that
sex-factors are involved with the others concerned in such a relationship as this we have a system of distribution approximating
to that found in *biennis* and *muricata*. The difference in reciprocals is represented in a not improbable way. It cannot yet
be said that the rarer terms in the series are formed at all, and
perhaps they are not. As we pointed out in our discussion of
these phenomena, the peculiar distribution of factors in these
cases must be taken to mean that the planes of division at some
critical stage in the segregation are determined with reference
to the parental groups of factors, or in other words, that the
whole system has a polarity, and that the distribution of factors
with reference to this polarity differs according to the grouping
of factors in the gametes which united in fertilization to produce
the plant. Subsequent proliferation of cells representing certain
combinations would then lead to excess of the gametes bearing
them. It is on similar lines that I anticipate we shall hereafter
find the interpretation of the curious facts discovered by de Vries,
though it is evident that a long course of experiment and analysis
must be carried through before any certainty is reached. The
work must be begun by a careful study of the descent of some
single factor, for example, that causing the broader leaf of
biennis, and we may hope that the study of *Oenothera* by proper
analytical methods will no longer be deferred.

We have now to return to the relations of *laeta* and *velutina*.
These two forms, it will be remembered are frequently produced
when *Lamarckiana* or one of its derivatives is used as male,
and the most unexpected feature in their behaviour is that *both
breed true as regards their essential characteristics, on self-fertilisation*. If one only bred true the case might, in view of the
approximate numerical equality of the two types, be difficult
to interpret on ordinary lines, but as both breed true it must be
clear that some quite special system of segregation is at work.
What this may be cannot be detected on the evidence, but with

the results from the *biennis-muricata* experiments before us, it is natural to suspect that we may here again have to recognise a process of allocation of different factors to the male and female sides in *laeta* and *velutina*. That some such system is in operation becomes the more probable from the new fact which de Vries states in describing the group of characters which he calls *conica*, namely that this type is the same as that of *velutina*.

There are many collateral observations recorded both by de Vries and others which have a bearing on the problems, but they do not yet fall into a coherent scheme. For example, we cannot yet represent the formation of *laeta* and *velutina* from the various species fertilised by *Lamarckiana* ♂. That this is not due to any special property associated with the pollen of *Lamarckiana* is shown by the fact that a species called *Hookeri* gives *laeta* and *velutina* in both its reciprocal crosses with *Lamarckiana* (de Vries, 1909, p. 3), and also by the similar fact that *Lamarckiana* ♀ fertilised by the pollen of a peculiar race of *biennis* named *biennis Chicago* throws the same types. Before these very complicated phenomena can be usefully discussed particulars must be provided as to the individuality of the various plants used. This criticism applies to much of the work which de Vries has lately published, for, as we now know familiarly, plants to which the same name applies can be quite different in genetic composition.

Attention should also be called to one curiously paradoxical series of results. When the dwarf "mutant" of *Lamarckiana* which de Vries names "*nanella*" is used as father on *muricata*, F₁ consists of *laeta* and *velutina* in approximately equal numbers. Both forms breed true to their special characteristics, but *velutina* throws dwarfs of its own type, while *laeta* does not throw dwarfs. Subsequent investigation of the properties of these types has led to some remarkable conclusions, and it was in a study of these plants that de Vries first came upon the phenomen of dissimilarity between the factors borne by the male and female cells of the same plant, a condition which had been recently detected in the Stocks as a result of Miss Saunders's investigations. The details are very remarkable. We have first the fact that *muricata* ♀ ×dwarf *nanella* ♂ gives about 50 per cent. *laeta* and about 50 per cent. of *velutina*.

As regards *Velutina* it was shown that:

		Talls, per cent.	Dwarfs, per cent.
1.	*Velutina* selfed gave	38	62
2.	*Velutina* ♀ × dwarf *nanella* ♂ gave	39	61
	do. × do. gave	49	51
	do. × dwarf ♂ derived from *velutina* gave	43	57
3.	Dwarfs × *velutina* ♂ gave	—	all dwarfs

The three experiments taken together prove, as de Vries says, that the ovules of *velutina* are mixed, talls and dwarfs, and that the pollen is all dwarf. The condition is almost the same as that of the Stocks. It may be noted also that in the Stocks the egg-cells of the "double" type are in excess, being approximately 9 to 7 of the "single" type, but de Vries regards the two types in *velutina* as probably equal in number. The figures (169:231) rather suggest some excess of the recessives, perhaps 9:7, and the point would be worth a further investigation.

As regards *laeta*, by self-fertilisation *no dwarfs were produced*, but in all other respects it behaved almost exactly like *velutina*. The ovules are evidently mixed talls and dwarfs, and whether fertilised by dwarfs or by the pollen of *velutina*, which is already proved to be all dwarf, the result was a steady 50 per cent. of talls and 50 per cent. of dwarfs. The pollen of *laeta* used on dwarfs gives nothing but dwarfs, and in three series of such experiments 226 dwarfs were produced.

We are thus faced with this difficulty. Since the egg-cells of *laeta* are evidently mixed, talls and dwarfs, and the pollen used on dwarfs gives all dwarfs, why does not self-fertilisation give a mixed result, talls and dwarfs, instead of *all talls*? De Vries regards the result of self-fertilisation as showing the real nature of the pollen, and declares it to be all talls, while he represents the behaviour of the same pollen used on dwarfs by stating that in these combinations the dwarf character dominates. This does not seem to me a natural interpretation. I should regard the pollen of *laeta* as identical with that of *velutina*, namely dwarf, and I suspect the difficulty is really created by the behaviour of *laeta* on self-fertilisation. Until a proper analysis is made in

9

which the identity of the different individuals used is recorded, no further discussion is possible.[8]

Other results of a complicated kind involving production of *laeta* and *velutina* together with a third form have been published by de Vries in his paper on "Triple Hybrids." To these also the same criticism applies. Some of the observations seem capable of simple factorial representation and others are conflicting.

Taking the work on *Oenothera* as a whole we see in it continually glimpses of order which further on are still blocked by difficulties and apparent inconsistencies. Through such a stage all the successful researches in complicated factorial analysis have passed and I see no reason for supposing that with the application of more stringent methods this more difficult set of problems will be found incapable of similar solutions. To return to the original question whether in *Oenothera* we can claim to see a special contemporaneous output of new species in actual process of creation, it will be obvious that while the interrelation of the several types is still so little understood, such a claim has no adequate support. It is true that many of the "mutants" of *Lamarckiana* can well pass for species, but this is equally true of many new combinations of pre-existing factors as we have seen in *Primula Sinensis* and other cases. Still less can it be admitted that these facts of uncertain import supply a justification for the conception which has played a prominent part in the scheme of the *Mutationstheorie*, namely that there are special periods of Mutation, when the parent-species has peculiar genetic properties. To conclude: The impression which the evidence leaves most definitely on the mind is that further discussion of the bearing which the *Oenotheras* may have on the problem of evolution should be postponed until we have before us the results of a searching analysis applied to a limited part of the field. In such an analysis it is to be especially remembered that we have now a new clue in the well-ascertained fact that the genetic composition of the male and female germ-cells of the

[8] Zeijlstra in a recent paper announces that many *nanella* plants are the subject of a bacterial disease to which he attributes their dwarfness. I gather that this does not apply to all *nanella* plants and that some are dwarfs apart from disease. The matter may no doubt be further complicated from this cause.

same individual may be quite different. When with this possibility in view the behaviour of the types is re-examined I anticipate that many of the difficulties will be removed.

Outside the evidence from *Oenothera*, which, as we have seen, is still ambiguous, I know no considerable body of facts favourable to that special view of Mutation which de Vries has promulgated. Of variation, or if we will, Mutation, in respect of some one character, or resulting from recombination, there is proof in abundance; but of that simultaneous variation in several independent respects to which de Vries especially attributes the origin of new specific types I know only casual records which have yet to undergo the process of criticism.

Besides de Vries's "*Mutationstheorie*" and "Species and Varieties" the chief publications relating to the subject of the behaviour of *Oenothera* are the following: (Many other papers relating especially to the cytology of the forms have appeared.)

Davis, B. M. Genetical Studies on *Oenothera*, I. *Amer. Nat.*, XLIV, 1910, p. 108.
Genetical Studies on *Oenothera*, II. *Ibid.*, XLV, 1911, p. 193.
Gates, R. R. An Analytical Key to some of the Segregates of *Oenothera*. *Twentieth Annual Report of the Missouri Botanical Garden*, 1909.
Studies on the Variability and Heritability of Pigmentation in *Oenothera*. *Ztsch. f. Abstammungslehre*, 1911, IV, p. 337.
Honing, J. A. Die Doppelnatur der *Oenothera Lamarckiana*. *Ztsch. f. Abstammungslehre*, 1911, IV, p. 227.
Macdougal, D. T. (with A. M. Vail, G. H. Shull, and J. K. Small). Mutants and Hybrids of the *Oenotheras*. *Carnegie Institution's Publication*, No. 24, 1905.
Macdougal, D. T., Vail, A. M., Shull, J. H. Mutations, Variations and Relationships of the *Oenotheras*. *Carnegie Institution's Publication*, No. 81, 1907.
de Vries, H. On Atavistic Variation in *Oenothera cruciata*. *Bull. Torrey Club*, 1903, Vol. 30, p. 75.
On Twin Hybrids, *Bot. Gaz.*, Vol. 44, 1907, p. 401.
Ueber die Zwillingsbastarde von *Oenothera nanella*. *Ber. Deut. Bot. Ges.*, 1908, XXVI, *a*, p. 667.
Bastarde von *Oenothera gigas*. *Ibid.*, p. 754.
On Triple Hybrids. *Bot. Gaz.*, 1909, Vol. 47, p. 1.
Ueb. doppeltreziproke Bastarde von *Oenothera biennis* L. und *Oenothera muricata* L. *Biol. Cbltt.*, 1911, XXXI, p. 97.
Zeijlstra, H. H. *Oenothera nanella* de Vries, eine krankhafte Pflanzenart. *Biol. Cbltt.*, 1911, XXXI, p. 129.

NOTE.

Since this chapter was written two contributions of special importance have been made to the study of the *Oenothera* problems. The first is that of Heribert-Nilsson.[9] The author begins by giving a critical account of the evidence for de Vries's interpretation of the nature of the mutants. In general this criticism pursues lines similar to those sketched in the foregoing chapter, concluding, as I have done, that the chief reason why factorial analysis has been declared to be inapplicable to the *Oenothera* mutants is because no one has hitherto set about this analysis in the right way. He has also himself made a valuable beginning of such an analysis and gives good evidential reasons for the belief that at least the red veining depends on a definite factor which also influences the size of certain parts of the plant. He argues further that many of the distinctions between the mutants are quantitative in nature. With great plausibility he suggests that the system of cumulative factors which Nilsson-Ehle discovered in the case of wheat (subsequently traced by East in regard to maize) may be operating also in these *Oenotheras*. According to this system several factors having similar powers may coexist in the same individual, and together produce a cumulative effect. Scope would thus be given for the production of the curious and seemingly irregular numbers so often recorded in the "mutating" families.

Another remarkable observation relating to the crosses of *muricata* and *biennis* has been published by Goldschmidt.[10] He finds that in the formation of this cross the female pronucleus takes no part in the development of the zygotic cell, but that when the male pronucleus enters, the female pronucleus is pushed aside and degenerates. As de Vries observed, the reciprocal hybrids are in each case very like the father ("*stark patroklin*"), a consequence which finds a natural explanation in the phenomenon witnessed by Goldschmidt. The results of the subsequent matings can also be readily interpreted on the same lines. Indications of maternal characters are nevertheless

[9] *Zts. f. Abstamm.*, 1912, VIII.
[10] *Arch. f. Zellforschung*, 1912, IX, p. 331.

mentioned by de Vries, and if Goldschmidt's account of the cytology is confirmed, these must presumably be referred to the influence of the maternal cytoplasm. Clearly this new work opens up lines of exceptional interest. The interpretation I have offered above must probably be reconsidered. The distinction between the male and female cells of the types may no doubt be ultimately factorial, but it is difficult to regard such a distinction as created by a differential distribution of the ordinary factors.

CHAPTER VI

VARIATION AND LOCALITY

In all discussions of the modes of Evolution the phenomena of Geographical Distribution have been admitted to be of paramount importance. First came the broad question, were the facts of distribution consistent with the Doctrine of Descent? I suppose all naturalists are now agreed that they are thus consistent, and that though some very curious and as yet inexplicable cases remain to be accounted for, the distribution of animal and plant life on the face of the earth is much what we might expect as a result of a process of descent with modification. Passing from this general admission to the more particular question whether the facts of distribution favour one special conception of the mode of progress of evolution rather than another, no agreement has yet been reached. One outstanding feature is hardly in dispute, namely that prolonged isolation is generally followed by greater or less change in the population isolated. Groups of individuals which from various causes are debarred from free intermixture with other groups almost always exhibit peculiarities, but on the other hand, cosmopolitan types which range over wide areas are on the whole uniform, or nearly so throughout their distribution. Examples of these two categories will be familiar to all naturalists. The barriers to intercourse may be seas, deserts, prairies, mountain-chains, or circumstances of a much less obvious character which isolate quite as effectually. The local unit is not necessarily an island, a district, or an area of special geological formation, but may, as every collector knows, be a valley, a pond, a creek, a "bank" in the sea, a clump of trees, a group of rocks in a bay, or a particular patch of ground on a mountain side. All the great groups provide examples of such specially isolated forms. The botanist knows them well; the conchologist, the entomologist, the ornithologist and the student of marine life are all equally aware that special varieties or special species come from special places and from nowhere else. In one remarkable case the season of appearance plainly acts as the isolating barrier. *Tephrosia bistortata* is a small Geometrid moth which has two broods, appearing in *March* and *July* respectively. It is closely allied to *T. crepuscularia* which emerges in *May* and *June*. From the fact that occasional specimens cannot be quite certainly referred to one or other of the two, many have held that the two are one species. Nevertheless, in general they present distinctions which are plain enough. Some localities have one form only, but in several woods they co-exist. Experiment has shown that the two can be crossed, and that the cross-breds can breed *inter se* and with at least one of the parent stocks.[1] Some diminution in fertility was observed, but perhaps not more than is commonly encountered when wild forms are bred in captivity. In such a case it can scarcely be doubted that the distinctness of the two forms in the places where they co-exist is maintained by the seasonal isolation.

Just as the consequences of isolation are to be seen in the most different forms of life so may they also affect the most diverse features of organisation, such as size, colour, sculpture, shape, or number of parts. In the Sloth (*Choloepus*) the geographical races differ in the number of cervical vertebrae—or in other words, in the distribution of vertebral differentiation. The geographical races of *Cistudo* differ in the number of claws and phalanges.[2]

In Shetland, the males of *Hepialus humuli* (the Ghost Moth) are not sharply differentiated in colour from the females, as they are elsewhere, but in varying degrees resemble them.[3] No such males are found in other localities, and even in the other Scottish islands they are normal. In the island of Waigiu the converse phenomenon has been observed in *Phalanger maculatus*. Gen-

---

[1] For the evidence see Tutt, J. W., *Trans. Ent. Soc.*, 1898, p. 17. Compare the remarkable case given by Gulick (*Evolution Racial and Habitudinal*, p. 123) of the two races of *Cicada*, which are separated by reason of their life-cycles, one having a period of 13, the other 17 years.

[2] For references see *Materials*, p. 396, and also G. Baur, *Amer. Nat.*, 1893, July, p. 677.

[3] Jenner Weir, *Entomologist*, 1880, XIII. p. 251.

erally the male is spotted with white, and the female is without spots, but in Waigiu the females are spotted like the males.[4]

The following striking illustration was pointed out to me by Dr. W. D. Miller. *Euphonia elegantissima* as it occurs in Mexico and Central America has the two sexes very distinct from each other. The male has the lower parts orange and the upper parts a dark indigo blue, with a bright turquoise-blue head and neck. The female, except for the head, is of a bright olive green. A form in which the sexes are similarly differentiated exists in Porto Rico and is known as *E. Sclateri.* But in many of the other West Indian islands the representative "species" (*E. flavifrons*) has the two sexes closely resembling the *female* of *E. elegantissima.* This form is found in Antigua, Barbados, St. Vincent, and Guadeloupe, from which localities the British Museum has specimens. All three so-called species are very much alike otherwise.

In the genus *Pyrrhulagra* (*Loxigilla*) to which Mr. Outram Bangs called my attention, several distinct and alternative possibilities occur. The genus has many local species occurring on the various West Indian islands. These species are characterized by differences in size, colour, and the shape of the bill. The colours have a narrow range, being black or greyish, with or without chestnut marks about the head and throat. In most of the islands the males are in general colour a full black, and the females are distinctly grey. They are thus found in San Domingo, Jamaica, Bahama, and most of the Lesser Antilles. In Porto Rico we meet the peculiarity that the hens are almost as black as the males (Ridgway describes the black of the hens as slightly less intense). This form is called *portoricensis.* A larger type, known as *grandis*, similarly coloured, inhabits St. Kitt's. Then, on the contrary, in Barbados, *both sexes* are a dull blackish grey, like the hens of the Lesser Antilles in general.

The local species of *Agelaius* show similarly capricious distinctions. *A. phoeniceus* is a widely spread species, found over a great part of North America. The male is black with red-orange

---

[4] Jentink, *Notes Leyden Mus.*, 1885, VII, p. 111. Specimens illustrating this peculiarity are in the British Museum.

bars on the wings, but the female is somewhat thrush-like in colour. In the island of Porto Rico there is a form called *xanthomus*, in which *both sexes* are like the males of the mainland. A similar species called *humeralis*, also with both sexes male-like, lives in Cuba. The island of Cuba, curiously enough, has also a distinct species named *assimilis*, in which the female is a dull black all over, though the male is like the mainland type.

So also may local races differ in respect of variability. *Argynnis paphia*, the Silver Washed Fritillary, through a great part of its distribution has only one female form. In the English New Forest a second female form, *valesina*, co-exists with the ordinary *paphia* female. But in the southern valleys of the Alps the *valesina* female is much the commoner of the two, and indeed in some localities where the species is abundant, I have seen no *paphia* females in many days collecting.

The beetle *Gonioctena variabilis* furnishes an illustration of a comparable phenomenon affecting the male sex. In 1894 and 1895 I studied the curious colour variations of this species especially in the neighbourhood of Granada, and Mr. Doncaster ten years later repeated the observations on the same ground, and also collected the insect in other places in the south of Spain. The distinctions are not easy to give in words and the reader is referred to the colour plate accompanying my paper.[5] The essential fact is that the males commonly have the elytra *red with black spots* and the females for the most part have greenish grey elytra with black stripes. In some localities a large minority of males closely resemble the female type, being identical in colour and then only distinguishable by structural differences. In two Granada localities I found the proportion of such males quite different. In the Darro valley about 38 per cent. (in

[5] *Proc. Zool. Soc.*, 1895, p. 850. Plate. Many points beyond that mentioned above are involved in this remarkable case. For example, not only are there males like females, but a small proportion of females resemble the ordinary male type. The stripes are not merely the spots produced, for they occupy different anatomical positions. The spots almost always go with a black ventral surface, but the striped forms nearly always have that region testaceous. *Spartium retama*, the food-plant, will not grow in England, but if it could be naturalised in America the whole problem might be investigated there and results of exceptional interest would almost certainly be attained.

718) were of this feminine type, but on the hills some 300 feet above only 19 per cent. (in 3,230) were like the females. At Castillejo, not far from Toledo I found no such male in 75 specimens.

Mr. Doncaster collected from several localities, especially from two areas near Malaga, about 5 miles apart. In one of these the female-like males were, as usual, in a minority, but in the other these were actually in great excess, amounting to about 81 per cent. in the 173 taken. Doncaster found a doubtful indication that the composition of the population varies with the season, which is quite possible, but it is most interesting to note that in my chief locality after the lapse of ten years he found the proportions very much the same as I had done at the same season, for where I had 19 per cent. of the female-like males his collecting gave 16 per cent. In other respects also, his statistics corresponded very closely with mine.[6]

The various forms of *Heliconius erato* are well known to entomologists. They are strikingly distinguished by the colours of the strong comb-like marking on the hind wing, which may be red, yellow, green or blue. In various parts of the distribution in South America sometimes two and sometimes three of these distinct types co-exist.[7]

The distribution of the varieties of *Noctua castanea* typifies a large range of cases. The form which is reckoned the normal of the species has red fore-wings. It is practically restricted to Great Britain and Germany, according to Tutt. The other common form, *neglecta*, has grey fore-wings, and in this pattern it ranges through West Central Europe from North Italy to Germany. In the British Isles it extends up to Orkney. In Britain this grey form is by far the commoner, occurring where-

[6] Doncaster, L., *Proc. Zool. Soc.*, 1905, II, p. 528.

[7] I am not aware that the details of this striking case have ever been worked out. It should be noted that the green and blue forms are not due to simple modification of the red pigment; for these colours, due to interference, fork over the area occupied by the red lines. The distinctions between these forms cannot therefore be simply chemical, as we may suppose them to be, for instance, in the case of many red and yellow forms, and the genetic relationships of the *Heliconid* varieties would raise many novel problems and be well worth studying experimentally.

ever the species is found. The red form is much scarcer in England, and does not occur at all in many localities where the grey form is common. Mr. Woodforde, from whom this account is taken,[8] states that in August, 1899, he saw considerably over a hundred of the grey in the New Forest at sugar, but only two red ones. In Staffordshire however the red is proportionately more numerous and he estimates them as 40 per cent. of the population. Lastly a form has been taken in Staffordshire as a rarity in which the red is replaced by yellow, and this has hitherto been seen nowhere else. It is beyond our immediate purposes to discuss the genetic relationships of such forms, but the details of this case are interesting as making fairly clear the fact that the distinctions between *castanea* and *neglecta* are due to combinations of the presence of and absence of two pairs of factors, of which one produces a red pigment in the ground colour of the forewing and the other irrorates the same region with black scales. Mr. Woodforde states that all intermediates exist, and that in Staffordshire the greys always have a pinkish tinge. The yellow is doubtless another recessive to the red.

Species which are uniform in some localities may be polymorphic in others. Such a phenomenon is well exemplified by the orchid *Aceras hircina*. Of this species distinct varieties had previously been known in Germany, but Gallé[9] has lately given a detailed account of a number of most diverse forms found growing in a district of Eastern France. Without reference to his plates it is impossible to give any adequate conception of the profusion of types which the flowers of the species there assume. In some the lip is elongated to many times its usual length, twisting and dividing in a fashion suggesting some of the strangest of the Tropical Orchids. In others the labellum and the lateral petals are all comparatively short and wide (Fig. 13). Intermediates, combining these qualities in various degrees, were abundant, and the condition of the species, which was the only representative of the genus in the locality, recalls the extreme polymorphism of many of the Noctuid Moths.

[8] Woodeforde, F. C., *Trans. North Staffordshire Field Club*, XXXV, 1901, Plate.

[9] E. Gallé, *Compte Rendus du Congres Internat. de Bot. a l'Expos. Univ.*, 1900, p. 112.

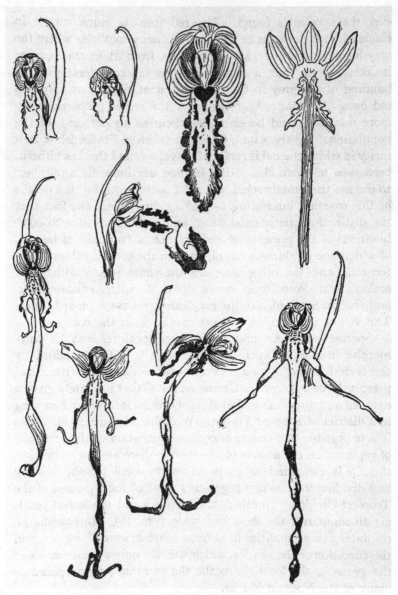

FIG. 13.   Various forms of *Aceras hircina*.   (After Gallé.)   This figure only shows
a few of the more striking forms illustrated in Gallé's plates.

Somewhat comparable variability has been seen in another Orchid genus *Ophrys*. In Great Britain the species *apifera*, *aranifera* and *muscifera* though variable are fairly distinct, but Moggridge has published two series of plates[10] showing a very different state of things as regards the *Ophrys* population of the Riviera. Here the outward diversity is such that the ordinary specific names cannot be applied with any confidence and the limits of the species are quite uncertain. It may well be supposed that these Riviera plants are interbreeding, and indeed we may safely assume that they are. It is, however, to be remembered that Darwin showed *apifera* in this country to be habitually self-fertilised, so that the different behaviour on the Riviera may itself constitute a local peculiarity. Moreover it is to be gathered from Moggridge's account that in the districts which he examined the condition was not to be described by the statement that our three types were there co-existing and hybridising, but rather we should say that the population was polymorphic, containing these three types amongst others. Conchologists are aware that on the Dogger Bank *Modiola* attains a size unparalleled elsewhere. The same is true of the sponges *Grantia compressa* and *Grantia ciliata* in the estuary of the Orwell.[11] Conversely, as we know so well in the case of Man, dwarf races occur in several special localities. Such examples may be multiplied indefinitely.

The relation of local forms to species has often been discussed from many points of view, but I know no treatment of the subject clearer or more comprehensive than an excellent account of some of the various manifestations of local differentiation as they appear in Helicidæ published by Coutagne[12] and a reader interested in the problem which they raise would

[10] Flora of Mentone, 1864–8, *Nova Acta Acad. Caes.*, XXXV, 1869.

[11] I owe these facts to Canon A. M. Norman, who showed me illustrative specimens. They were originally described by Bowerbank (*Monogr. Brit. Spongiadae*, vol. II, pp. 18 and XX; vol. III, Pls. I and III). A specimen of *G. compressa* measured 5 inches, with a greatest width of 3¼ in. *G. ciliata* was found measuring 3 in. long and ¾ in. wide. These dimensions are many times those of normal specimens.

[12] Coutagne, G., *Recherches sur le Polymorphisme des Mollusques de France*. *Annales Soc. d'Agric. Sci. et Industr. Lyon*, 1895.

do well to make himself acquainted with the original from which the following notes are taken. He speaks for example of *Helix lapicida*. This is on the whole a constant form ranging up to the altitude of 1,300 m., common all over France except at great heights and in the Olive regions where it is restricted to moist places. Though subjected to such diverse conditions it shows only trivial variations in colour and other respects throughout its distribution, excepting that on both sides of the Pyrenees it has a very distinct sporadic variety called *Andorrica* or *microporus*. This variety occurs here and there, together with the type-form sometimes in colonies (pp. 26–30 and 86).

Bulimus detritus though more restricted in geographical range is a much more variable form. It exhibits great variations in colour, form, and size, and as Coutagne well insists, these are independent of each other. Foreshadowing the methods of factorial analysis he suggests that distinctions in each respect, the "modes" as he calls them, should be denoted by a letter, or if desired, by a name, and the several combinations of differ-ences might thus be most logically and usefully expressed. Of such combinations he says there are at least 18, all of which can be found. The whole possible series does not necessarily occur in the same place, and various localities are characterised by the presence or absence of certain of the combinations as Coutagne calls them, and by the relative frequency with which they occur. The ideas thus enunciated are much in advance of the ordinary practice of systematists, who give names to forms which are nothing but accidental combinations of factors, just as the horticulturists for practical reasons give names to similar com-binations, which as we now know are merely specially noticeable terms in a long series of possibilities. In each case it is rather the *factors* which should be named than the forms which are constituted by their casual collocation. In this special example of *Bulimus detritus* the 18 forms are made by the combinations of three pairs of independent factors. Besides these combina-tions which may occur anywhere or almost anywhere in the dis-tribution there are two more distinct local forms, each of which is regarded by Coutagne as probably constituting a fresh " mode," perhaps compatible with the others.

Helix striata (Draparnauld)[13] is truly polymorphic; and its various forms have been described under various specific names. It abounds in the calcareous hills of Provence and Languedoc, disappearing in the alluvial lowlands and equally in the upper levels at about 800–1,000 m. From this district it extends through regions of similar altitude over a great part of France (details given).

Locard in his monograph of this group, which he calls collectively the group of *Helix Heripensis*, tabulates 27 distinct named forms. The characteristics in which these forms differ have been reckoned as 17, and as several of these vary in degree of development, the number of modes may be increased to 109. For practical purposes however Coutagne considers that the various developments of 7 characteristics in their several combinations are enough to express the various forms, and he gives examples of this method of definition. As he observes, though names may be required to define the modes, no one need be alarmed at that, for the same names of modes will be applicable to a great range of distinct species, and the formulae expressing their combinations will replace the varietal names.

This particular example of polymorphism is but little limited by locality. Occasional colonies present some special physiognomy which may in a given place seem almost invariable, though in this very respect the colonies found elsewhere may be highly variable, but such limitations are exceptional for *H. striata*.

Some distinct and obvious susceptibilities to the influence of soil and climate are however noticeable. For example on siliceous ground the shells are thinner, while on calcareous soils they are thicker; similarly those from the Northern districts attain a larger size than those from further South. Moreover those subjected to curtailed development, whether from drought, heat or cold often show a shortening of the spire. In contrast with this case Coutagne describes the varieties of *Helix caespitum*, which he says are for the most part localised, quoting many illustrative cases.

Another remarkable case in which locality plays a curious part is provided by the two species *Helix trochoides* and *pyra-*

---

[13] As to the synonymy and references see Coutagne, p. 45.

midata. In France generally they are distinct enough from each other, *trochoides* being smaller and having a characteristic keel. Coutagne says that after having collected these species from more than a score of localities he came upon a colony of *trochoides* on the island of Pomègues in which the shells were relatively enormous, most of them having only a slight keel, and a few none at all. On the other hand he received a consignment of *pyramidata* from four localities in Sicily, all small, and one of them exactly like the *trochoides* from Pomègues. Judging by the samples received from Sicily, *trochoides* is there not more variable than it is in Provence, while the Sicilian *pyramidata* is protean.

The relations of the two species *Helix nemoralis* and *hortensis* provide an illustration of another kind of manifestation of local peculiarity. *H. hortensis* and *nemoralis* as usually met with, are two very distinct forms. *H. hortensis* is smaller and duller, and its peristome is white. *H. nemoralis* is larger and more shiny, and its peristome is brown. In several anatomical points, moreover, especially in the shape of the dart, there are great differences. For a full account of these peculiarities of the two forms and a discussion of their inter-relations the reader is referred to the elaborate work of A. Lang[14] who has studied them extensively and has also succeeded in experimentally raising hybrids between them. These hybrids were in a slight degree fertile with both the parent species, but up to the time of publication no young had been reared from hybrids *inter se*.

Coutagne describes the result of collections made in 62 French localities. Some had exclusively *hortensis*, some exclusively *nemoralis*, and in some the two were found in association. He gives details of five of these collections from which I take the following summary of the more essential facts, omitting much that is almost equally significant.

Locality A, near Honfleur. Both forms present, each sharply and normally distinguished, without any intermediates. They

[14] A. Lang, *Die Bastarde von H. hortensis Muller H. nemoralis L.* Jena, G. Fischer, 1908; with a fine coloured plate showing the varieties of the species and their hybrids.

are thus found in many places. Coutagne instances Müller's observations in Denmark, his own series from the Jura, etc.

Locality B. Vonges (Côte d'Or), 242 *hortensis* taken at random, showed 128 with light peristomes (either more or less pinkish or quite white) and 114 with dark *brown* peristomes; together with 26 *nemoralis* all with the usual brown peristomes.

Of the *hortensis* 50 were in ground-colour *opalescens* and 1 *roseus;* and in shape 5 were *umbilicatus.*

Locality C, about 3 kilometres from *B.* There were found 35 *hortensis,* of which 20 had light peristomes and 15 brown; together with 7 *nemoralis.*

Of the *hortensis* none were *opalescens;* 18 were *roseus* and none has the shape of *umbilicatus.*

Locality D, about 1,200 metres from *B.* 147 *hortensis,* of which 4 had light peristomes and 143 had brown. No *nemoralis* were found.

None of the *hortensis* were *opalescens* or *roseus,* but 30 were *umbilicatus.*

In these localities intermediates of every grade existed between the well-characterised *opalescens, roseus,* or *umbilicatus,* and the other forms, but there were no intergrades between the other *nemoralis* and the smaller *hortensis,* about which there was no hesitation. In the next locality a very different state of things was found.

Locality E. Banks of the Yvette at Orsay (Seine-et-Oise). The actual numbers are not given, but we are told that 58 per cent. were *hortensis,* 33 per cent. *nemoralis,* and 9 per cent. intermediate. As at Honfleur, the *hortensis* had white peristomes, and the *nemoralis* brown. Coutagne's visits to this locality were in 1878 and 1880, and he calls attention to the fact that Pascal found similar intermediates in the same neighbourhood in 1873.

The two species, in Coutagne's view, when they occur together, can generally be sorted from each other with perfect confidence, and it is only in exceptional localities that these intermediates occur. Whether they are hybrids, or whether sometimes the species in their variations transgress their usual limitations is regarded both by Coutagne and by Lang as a

question not yet answerable with certainty. Coutagne moreover
lays stress on the fact that although each species may be easily
known from the other *in its own district*, yet when shells from
different districts are brought together it is sometimes impossible
to sort them. He mentions an example of such casual intermixture occurring under natural conditions on an island in the
Rhone, to which it may well be supposed that floods had brought
immigrants from miscellaneous localities. This population contained a very large number of uncertain specimens, and as he
says, it was much as if he were to mix the shells from his 62 localities, after which it would certainly be impossible to separate
the two species again.[15]

Further evidence is given in the same treatise as to other
examples of polymorphism, especially in the genus *Anodonta*, of
which Locard made 251 species for France alone. Here again
are cases like those already given, and many forms or "modes"
are found restricted to special localities, while occasionally
in the same locality dissimilar forms are found, collectively
forming a colony, without intermediates.

Taken as a whole the evidence shows the following conclusions
to be true. Local races, whether of animals or plants, may be
distinguished by characters which we are compelled to regard
as trivial, or again by features of such magnitude that if they
were known to us only as the characteristics of a uniform species
they would certainly be assumed without hesitation to be essential
for its maintenance. Local forms may be sharply differentiated
from the corresponding populations of other localities or they
may be connected with them by numbers of intermediates.
Not rarely also we find a fact which has always seemed to me of
special significance, that the peculiarity of the local population
or colony may show itself in a special liability to variation, and
this variability may show itself in one of many degrees, either
in the constant possession of a definite aberration, in a dimor-
phism, or in an extreme polymorphism.

At this stage attention should be called to two points.

[15] With this evidence compare that given by A. Delcourt in his valuable
papers lately published relating to the variations of *Notonecta*. See especially
Bull. Sci. Fr. Belg., 1909, XLIII, p. 443; and *C. R. Soc. Biol.*, 1909, LXVI, p. 589.

First, that when the details of the geographical distribution of any variable species are studied in that thorough and minute fashion which is necessary for any true knowledge of the inter-relations of the several forms, the conception of a species invented by the popular expositions of Evolution under Selection is found to be rarely if ever realised in nature.

A species in this generalised sense is an aggregate of individuals, none exactly alike, but varying round a normal type, the characters of which are fixed in so far as they are adapted to environmental exigency. In nature, however, the occurrence of the varieties, and even the occurrence of the variability is sporadic. In one place a population may be perfectly uniform. In another it may be again uniform but distinct. In others the two forms may occur together, sometimes with and sometimes without intergrades. In some localities a sporadic variety may be an element of the population, persisting through long periods of time. In other localities there may be several such aberrations occurring together which are absent elsewhere.

Secondly, I would remind the reader that in the light of genetic analysis we know that intergrades, when they do occur, cannot be assumed to represent conditions through which the species must pass or has passed on its way to the extreme and definite forms.

Often, perhaps generally, they are nothing but heterozygous forms, and often also they are conditions corresponding with the presence of factors in their reduction-stages.

A broad survey of the facts shows beyond question that it is impossible to reconcile the mode of distribution of local forms with any belief that they are on the whole adaptational. Their peculiarities are occasionally the result of direct environmental influence, as we shall hereafter notice in certain cases, but none can attribute such sporadic and irregular phenomena to causes uniformly acting.

Writers on systematics, especially those of former generations often conjecture or assert that local distinctions are caused by "differences of climate, soil, food, etc.," in vague general terms. It is usually safe to assume that these remarks do not represent conclusions drawn from actual evidence, for only rarely can they be translated into more precise language. So thoroughly have the biological sciences become permeated with the belief that all distinctions are dependent upon adaptation, that the mere existence of definite distinctions is felt by many to be sufficient ground to warrant an assumption that these distinctions are directly or indirectly due to special local conditions. For example, Dr. J. A. Allen, who has done so much careful and valuable work in delimiting the local forms of the United States fauna, writes of the Ground Squirrels (Tamias)[16] as follows:—

"From the extreme susceptibility of this plastic group to the influences of environment, it is one of the most instructive and fascinating groups among North American mammals. No one can doubt its comparatively recent differentiation from a common stock, and its dispersion from some common centre. Whether the type originated at some point in North America, or in the Northern part of Eurasia, it is perhaps idle to speculate, but that it has increased, multiplied, spread, and become differentiated to a wonderful degree in North America is beyond question; as it is found from the Arctic regions to the high mountain ranges of Central Mexico, and has developed some twenty to thirty very palpable local phases."

"Some of them easily take rank as species, others as subspecies. Probably a more striking illustration of evolution by environment cannot be cited."

He proceeds to point out that the habits of these creatures are such as lead to isolation. This may well be admitted, and indeed no exception can possibly be taken to the passage as a whole, save in the one respect that there is no real proof that the local diversity is due to "evolution by environment" or an indication of "susceptibility to the influences of environment."

Dr. Allen does indeed adduce the fact that California "extending through 800 miles of latitude, with numerous sharply contrasted physiographic regions, has apparently no less than six strongly differentiated forms, while the region east of the Rocky Mountains from a little below the northern boundary of

[16] Allen, J. A., *Bull. Amer. Mus. N. H.*, III, 1891, pp. 51–54.

the United States northward to the limit of trees—a slightly diversified region of at least ten times the area of California—has only one"! But when one comes to ask how the various forms are adaptational, and how the influences of environment have led to their production, only conjectures of a preliminary and tentative character could be expected in reply. Desert forms are no doubt pallid as in so many instances, and forest forms are more fully coloured, and we may readily enough accept such facts as indications of a connection between bodily features and the conditions of life, but further than that no one can go; so that when we find size, length of ears or of tail, the number of dorsal stripes, the pattern of the colours, not to speak of differences in the pigments themselves, all exhibiting large modifications, we cannot refer these peculiarities to the causation of environmental difference, save as a simple expression of faith. I incline far more to agree with Gulick who, after years of study of the local variations of the Achatinellidae, came to the conclusion that it was useless to expect that such local differentiation can be referred to adaptation in any sense.[17] Even the most convinced Selectionist must hesitate before such facts as those related by A. G. Mayer regarding the distribution of *Partula otaheitana*, one of these Achatinellidae. The island of Tahiti has been scored by erosion so that a series of separated valleys radiate to the coast. From four successive valleys Mayer collected the species, and found that in the first (Tipaerui) valley all the shells were dextral (115, containing 73 young); in the second valley (Fautaua) 54 per cent. of adults and 55.5 per cent. of the young' contained were sinistral; in the third valley (Hamuta) 69 per cent. of adults and 73 per cent. of young contained in them were sinistral; and lastly, in the fourth valley (Pirae) all the shells (131, containing 62 young) were sinistral.[18] In connection with these observations I may mention the fact that in a certain pond in the North of England[19] the sinistral form of *Limnaea*

[17] J. T. Gulick, *Evolution, Racial and Habitudinal*, Carnegie Institution, Publication No. 25, 1905.

[18] A. G. Mayer, *Mem. Mus. Comp. Anat. Harvard*, Vol. XXVI, 1902, p. 117. From the tables given I cannot ascertain the actual numbers from the two intermediate valleys, but they were considerable.

[19] To which I was very kindly guided by Mr. C. T. Trechmann.

peregra has been known to occur for about fifty years. Visiting it lately I found the left-handed shells to be about 3 per cent. of the population. The species is the commonest British fresh-water shell, but left-handed specimens are exceedingly rare. Will anyone ask us to suppose that the persistence of a percentage of this rarity in the same place is an indication of some specially favouring circumstance in the waters of that pond? It is a horse-pond to all appearances exactly like any other horse-pond; and I believe that in perfect confidence we may accept the suggestion of common sense, which teaches us that there is nothing particular in the circumstances which either calls such varieties into existence or contributes in any direct way to their survival. Had the phenomenon of local variation been studied in detail before Darwin wrote, the attempt to make selection responsible for fixity wherever found, could never have been made. The proposition that not only the definiteness of local forms but their variability also is sporadic, can be established by countless illustrations taken from any group of either the animal or the vegetable kingdoms. Only exceptionally can the fixed differences be even suspected of contributing to adaptation, and sporadic variability, which is a no less positive fact, must manifestly lie outside the range of such suspicions. It is open to any one to suggest speculatively that the persistence of special varieties or of special variability in special places is an indication that in those places the conditions of life are such that the forms in question are tolerated though elsewhere the same types are exterminated; but that consideration, even if it could be proved to be well founded, is not one which lends much force to the thesis that definiteness of type is a consequence of Natural Selection. On the contrary, recourse to such reasoning implies the inevitable but very damaging admission that the stringency of Selection is frequently so far relaxed that two or more equally definite forms of the same species can persist side by side. There is no doubt that this is the simple truth, but when once that truth is perceived it is useless to invoke the control of Selection as the factor to which definiteness of type in general must be referred.

The genetic relations of local forms to each other cannot in the absence of actual breeding experiments be often ascertained. Standfuss formerly enunciated as a general principle that when two forms co-exist in the same locality and are able to interbreed, they do not produce intermediates; but that when the forms are geographically separated as local races, crosses between them result in a series of intermediates.[20]   In this aphorism there is a good deal of truth, but if in the light of Mendelian principles we examine the two statements we see now that the first is in reality only another way of saying that the distinctness of an aberrational form co-existing with another is due to segregation, accompanied by some degree of dominance of one type.   Whether, however, one geographically isolated race will give intermediates when bred with another must depend entirely on the genetic physiology of the special case, and no general rule can be laid down.   It may well be that, inasmuch as the distinctness of the variety is maintained by isolation, the difference in factorial composition between it and the representative form in another area is neither simple nor sharp; but when two varieties co-exist, though interbreeding, it is now clear that their differences must depend on the segregation of simple factors.   Plainly such aberrations may in one place co-exist with another type, and elsewhere be separated from it as local races.

Excellent illustrations of these two stages in evolution are provided by the melanic varieties of British Lepidoptera.   The fact that black or blackish varieties of many species especially of Geometridae have come into existence in recent years is well known to British collectors, and it is not in dispute that they have in several instances replaced the older type more or less completely in certain districts.   In the year 1900 the Evolution Committee of the Royal Society instituted a collective inquiry as to the contemporary distribution of these dark varieties.   As the change had happened within living memory and had greatly progressed in recent years it was hoped that a record of the existing distribution would serve as a point of departure for future comparison.   The records thus obtained were tabulated

[20] Standfuss, *Handbuch d. paläarkt Gross-schmet*, 1896, p. 321.

by Mr. L. Doncaster.[21] From that account and from the statements n Barrett's British Lepidoptera[22] this description of some
of the more notable cases is taken.

The most striking and familiar case is that of *Amphidasys
betularia*, of which only the ordinary type was known in any
locality until about 1848–1850, when the totally black var.
doubledayaria first appeared in the neighbourhood of Manchester.
This black form was subsequently recorded in Huddersfield
between 1860 and 1870; Kendal about 1870; Cannock Chase,
1878; Berkshire, 1885; Norfolk, Essex and Cambridge about
1892; Suffolk, 1894; London, 1897. For the Southern Counties
of England, except in the London district, there are still very few
records. It cannot of course be asserted positively that the
variety spread from its place of first appearance into the other
localities, and that it did not arise *de novo* in them, but there
can be little doubt that the process was one of colonisation.
On the European Continent the first records are from Hanover
in 1884, Belgium 1886 and 1894, Crefeld 188–, Berlin 1903,
Dresden about the same date.

As regards the increase of the variety we have the fact that
in Lancashire, Cheshire and the West Riding of Yorkshire the
black is now the prevalent form; and in some places, as for example,
Huddersfield, the black alone is now found, though it was unknown there till between 1860 and 1870. About 1870 at New-
port, Monmouth, the two forms were in about equal numbers,
but a few years later the type had almost vanished. Similarly
in Crefeld, where the black form was still very rare in the eighties,
it now forms about 50 per cent. of the population. In the
London district the black remains scarce and at the date of the
report it was still very scarce. From Ireland there is only one
record and there are hardly any from Scotland.

Boarmia repandata is another species which is behaving in a
somewhat similar way. Unlike *betularia*, however, the species
is a variable one, and has several colour-forms, amongst them
the banded var. *conversaria*, and many others. In addition

[21] *Ent. Rec.*, XVIII. No. 7, 1906.

[22] This evidence was largely collected by Mr. G. T. Porritt, who has given
much attention to the subject.

to these there is a black form in the North of England which seems to be spreading. In Huddersfield the black was first recorded in 1888, and in 1900 20–25 per cent. were black. At Rotherham the black or very dark are now prevalent and have increased in the last 15 years. From the Midlands, East Anglia and Southern Counties the returns show only the light and medium forms.

Of *Odontoptera bidentata* several intergrading dark forms exist, and these are found exclusively in the North and the Midlands. Unicolorous blacks have been found recently in the Lancashire mosses and at Wakefield. At Huddersfield 50 years ago the light forms were prevalent, but now a rather dark brown, not infrequently suffused with black, is the commonest. In Southern Counties only light forms are known.

Phigalia pilosaria in South England is always light, but in the North the prevalent form is darker. About 35 years ago a form with unicolorous sooty fore-wings and dull grey hind wings was first seen in Yorkshire and a similar form is now taken regularly in South Wales.

In the following cases the dark varieties were found originally only in the South.

Boarmia rhomboidaria gave rise about 40 years ago to a uni-colorous smoky variety called *perfumaria*. This was at first peculiar to the London district, but it has since been taken in Birmingham and other large cities. More lately coal-black specimens have been found at Norwich, and others similar but hardly so dark were taken in the South of Scotland and at Cannock Chase.

Eupithecia rectangulata is a similar case. Formerly the light forms were prevalent but within sixty years they have almost entirely been replaced in the South of London by a nearly black form.

Tephrosia (Boarmia) consortaria and *Tephrosia consonaria* are exceptionally interesting, for they have both given off dark forms in the same wood near Maidstone, which is far from the usual "centres of melanism." They were discovered in this locality by Mr. E. Goodwin. That of *consortaria* is a dark grey, but that of *consonaria* is a full black, and nothing like either has been found anywhere else.

These examples are all taken from the Geometridae but others, though of a less conspicuous kind, could be given from the Noctuidae or the Micro-Lepidoptera. *Acronycta psi*, for instance, has a suffused form which is believed to be becoming more frequent in the London district. *Polia chi* has two dark forms, *olivacea*, a yellowish grey with dark markings, and *suffusa* which is a darker, blackish-slate colour. Both occur in the North of England, sometimes together, sometimes separately, or mixed with the type and many intermediates. The distribution is peculiarly irregular. At Huddersfield, where the very dark form appeared suddenly about 1890, some 30 per cent. are said to be now dark and about 6–7 per cent. very dark, but at Saddleworth, 12 miles away, only the pale forms occur.

Several questions of interest arise in regard to this evidence. This progressive Melanism has arisen in certain families only, and may be confined to certain species only, within those families. As in almost all other examples in which variation has been much observed, its incidence is capricious and specific. A collateral line of inquiry relates to the degree of discontinuity which the variation manifests. Here again there is no rule. Generally speaking, in *A. betularia*, to take the case most fully studied, the variation is discontinuous. Real intermediates between *betularia* and *doubledayaria* are in most localities absent or rare. The black spots of *betularia* may often be larger or more numerous than in the normal, but this variation has nothing to do with *doubledayaria*, and is not an intermediate stage towards it, though sometimes wrongly so described. *Doubledayaria* owes its characteristic appearance to a factor which blurs the surface of the wings with a layer of black. Sometimes this blurring is slighter than in the real *doubledayaria*, and these forms are real intermediates. Occasionally the forewings alone are thus blurred. These intermediates are clearly due to reduction-stages of the *doubledayaria* factor, and are related to it as a blue mouse is to a black, or a dutch rabbit to a self-colour. It cannot positively be asserted that the full *doubledayaria* existed before the intermediate, but it almost certainly did. In certain places as for instance in Belgium, there is evidence that intermediates have at various times been fairly abundant, but they have never become common, nor are they known to exist in the absence of *doubledayaria*. When the black variety and the light type breed together they do not usually have intermediates among their offspring, and the evidence is consistent with the view that the black is a complete dominant. The same is probably true of *Tephrosia consonaria*.

In some of the other species we know that the darkest forms did not appear first. For example in *Phigalia pilosaria* and *Boarmia rhomboidaria* dark forms existed and are believed to have increased in number before the darkest made its appearance. *Hybernia progemmaria* is said to have become darker gradually both in Cheshire and in the West Riding, and a uniformly smoky variety appeared in South Yorkshire less than 45 years ago which has spread to neighbouring counties. The dark medium has become the commonest form in Huddersfield district, where the very dark variety is now about 20 per cent. of the population, though the light form is still common.

Taking the evidence together we find it consistent with the view that dark forms have appeared sporadically, in some species the very dark appearing first and intermediates later, in others the moderately dark came first and the darkest later in time. It is practically certain that the change has in general come about not by a gradual change supervening on the population at large, but by the sporadic appearance of dark specimens as a new element in the population, and strains derived from these dark individuals have gradually superseded the normal type more or less completely.

If it could be shown that these melanic novelties had a definite advantage in the struggle for existence they would provide an instance of evolution proceeding much in the way which Darwin contemplated. The whole process would differ from that conceived by him as the normal method of evolution only in so far as the change has come about with great rapidity and in some instances largely by the appearance and success of dis-

continuous varieties. The question, however, must be asked
whether the dark form can reasonably be supposed to have
an advantage by reason of their darkness. Some naturalists
believe that the darkness of the colours does thus definitely contribute to their protection by making the insects less conspicuous
and thus more likely to escape the search of birds. In support
of this view it may be pointed out that it is in the manufacturing
districts of Lancashire and Yorkshire, and again in the London
area that the melanics have attained their greatest development.
Consistently with this argument also, it is in the neighbourhood of
Crefeld and Essen, the black country of Germany, that they have
chiefly established themselves on the Continent, and *Phigalia
pilosaria* in the black form is now at home in South Wales. Thus
superficially regarded, the evidence looks rather strong, but it
is difficult to apply the reasoning in detail. We have first the
difficulty that the black form of *betularia* for instance has established itself in thoroughly rural districts, notably near King's
Lynn in Norfolk, and in the neighbourhood of Kendal and
Windermere. The black form of *consonaria* and the dark
consortaria appeared in a wood near Maidstone, far from town
smoke, and the black *rhomboidaria* was first found at Norwich,
which, as towns go, is clean. Then again the spread of the
melanics is very irregular and unaccountable. The black *pilosaria* is found both in the West Riding and in the Swansea
district, but not yet elsewhere. It rapidly increased at Huddersfield, but made no noticeable progress at Sheffield though recorded there for ten years. It is also a remarkable fact that no
similar melanic development has been observed in America,
and, so far as I am aware, comparable melanic varieties have not
appeared on the European continent except in the case of the
few sorts which possibly may have come from England.

The whole subject is beset with complications. It must
not be forgotten that in a few species of moths there is an obvious
and recognised conformity between the colours of the perfect
insect and that of the soil on which they live, comparable with
that which is so striking in the case of some Oedipodidae and
other grasshoppers. Of this phenomenon the clearest example

is *Gnophos obscurata*, which is a most variable species with many local forms. Of these a well-known dark variety lives on the peaty heaths of the New Forest and other districts, but on the chalk hills of Kent, Sussex and Surrey various light varieties are found, of which one is a bright silvery white, very near in colour to the colour of a chalky bank. This case does not seem to be one of direct environmental action,[23] for Poulton found no change induced by rearing larvae among either white or black surrounding objects. No one however can doubt that there is some indirect connection between the colour of the ground and that of the moths.

To my mind there is a serious objection to the theory of protective resemblance in application to such a case as that of the *betularia* forms, which arises from the fact that the black *doubledayaria* is a fairly conspicuous insect anywhere except perhaps on actually black materials, which are not common in any locality. Tree trunks and walls are dirty in smoky districts but they are not often black, and I doubt whether in the neighbourhood of Rotherham, for instance, which is one of the great melanic centres, *doubledayaria* can be harder for a bird to find than *betularia* would be. After all, too, many of the species much affected are not urban insects. They live in country places between the towns, and the general tone of these places even in Lancashire and the West Riding is not very different from that of similar places elsewhere. As against the objection that the black varieties are much blacker than the case requires it may be replied that we know nothing of the senses of birds, and that perhaps to their eyes blackness does constitute a disguise even though the surroundings are much less dark. This is undeniable, but recourse to such an argument is dangerous; for if the sight of the insect-eating birds is so dull that it does not distinguish dark things from dingy grey, we cannot subsequently regard the keen sight of birds as the sufficient control which has led to the minute and detailed resemblance of many insects to their surroundings. Those who see in such cases examples of

[23] Such direct action has of course been proved to occur in the case of several dimorphic larvae (*e. g., A. betularia*, itself) and pupae.

the omnipotence of Selection must frequently find themselves in this dilemma.

Taking the evidence as a whole, we may say that it fairly suggests the existence of some connection between modern urban developments and the appearance and rise of the melanic varieties. More than that we cannot yet affirm. It is a subject in which problems open up on every side, and all of them are profitable subjects for investigation. Unhappily such animals are difficult to rear successfully in captivity for many generations, owing to their extreme liability to disease. Not the least interesting feature of the melanics is the fact that the black varieties provide about the best and clearest example of a new dominant factor attaching itself to a wild species in recent times. None of the cases are satisfactorily recorded or analysed as yet, but the evidence is clear that *doubledayaria* is a dominant to its type, and in several other dark varieties, though the pigment deposited is not black, the records show that the increased amount of the pigment almost certainly is due to a positive factor. Of this, *Hemerophila abruptaria* is a good example.[24] There are some irregularities in the results, but taken together they leave little doubt that the dark brown variety is a dominant and the light, yellowish brown a recessive.

A curious parallel to the rise of the melanic moths in England is provided by the case of the Honey-creepers or Sugar-birds, in certain West Indian islands.[25] These birds of the genus *Coereba (Certhiola)* range from Southern Mexico to the Northern parts of South America and through the whole chain of the West Indian islands and Bahamas except Cuba. There are numerous local forms, and many of the islands have types peculiar to themselves, as is usual in such cases. Some of the types or species range through several islands, but according to Austin Clark[26] no island has more than one of them. Cory[27] reckoned twelve

[24] See Harris, *Proc. Ent. Soc. London*, 1904, p. lxxii, and 1905, p. lxiii; also Hamling, *Trans. City of London Ent. Soc.*, 1905, p. 5.

[25] I am indebted to Mr. Outram Bangs of the Harvard Museum for calling my attention to this remarkable case.

[26] *Auk*, 1889, VI, p. 219.

[27] *Ann. N. Y. Acad. Sci.*, 1878, I, p. 149.

such species within the Antillean region. They are small birds about the size of a nuthatch with a general colouring of black, yellow, and white. From the island of St. Vincent the Smithsonian Institution received in the late seventies of last century several completely black specimens in addition to two of the usual type of colouring. The black were described by W. N. Lawrence as *atrata*, and those marked with the usual yellow and white were called *saccharina*. The collector (Mr. F. A. Ober) reported that the black form was common, and that the *saccharina* form was rarer. Lawrence remarks, "Had there been only a single example (of the black form) I should have considered it as probably a case of abnormal colouring, but it seems to be a representative form of the genus in this island." [28] There is of course no doubt of the correctness of the view taken by Austin Clark that "*atrata*" is a black variety. The black bird is in every respect, other than colour, identical with *saccharina*, and it is even possible to detect a greenish colour in the areas which would normally be yellow, showing plainly enough the yellow pigment obscured by the black.

We have next the interesting fact that like our melanic moths the dark form is replacing the "type." At the time of Ober's visit the type was already in a minority, but now it is nearly or perhaps actually extinct, though the black form is one of the commonest birds on the island. Austin Clark found no specimen when he collected there in 1903–4, though formerly it was not uncommon in the vicinity of Kingston and in the immediate windward district of St. Vincent.

The Grenadines are geographically just south of St. Vincent, though separated by a deep channel. In these islands no black forms have yet been taken, but Grenada, the next island to the south, has both normals and blacks. There are trifling differences of size between the Grenada birds and those from St. Vincent, the Grenada specimens being slightly smaller and for this reason they have received distinct names, the form marked with yellow and white being called *Godmani* (Cory) and the black, *Wellsi* (Cory), but this merely introduces a useful complication.

[28] *Ann. N. Y. Acad. Sci.*, 1878, I, p. 149.

There is evidence that in Grenada, as in St. Vincent, the black is gradually ousting the original type, but the process has not gone so far as in St. Vincent. Austin Clark very properly compares this case of the Sugar-birds with that of *Papilio turnus*, which as is well-known, has a black female in the southern parts of its distribution, in addition to a female of the yellow type, but in the Northern States the black female does not occur.

During the present year P. R. Lowe, who lately studied *Coerebas* on a large scale in the West Indies, has published an important paper on the subject.[29] He calls attention to the fact that Cory recently found a black form of *Coereba* on Los Roques Islands, and he himself discovered another on the Testigos Islands. Both localities are on the coast of Venezuela, far from St. Vincent and Grenada. The whole problem is thus further complicated by the fact that the black varieties have, as we are almost driven to admit, arisen independently in remote places. Improbable as this conclusion may be, it is still more difficult to regard all the black forms as derived from one source. For first, they present definite small differences from each other; and secondly we have to remember a consideration of greater importance, that the very fact that each island has its own type must be accepted as proving that the localities are effectively isolated from each other, and that migration must be a very rare event.

The rarity of such illustrative cases is, I believe, more apparent than real. It is probably due to the extreme reluctance of systematists to admit that such things can be, and of course to the almost complete absence of knowledge as to the genetic behaviour of wild animals and plants. Only in such examples as this of the *Coereba*, where colour constitutes the sole difference, or that of the moths which have been minutely studied by many collectors, does the significance of the facts appear. The arrangement of catalogues and collections is such that much practical difficulty of a quite unnecessary kind is introduced. For example, in this very case of *Coereba*, I find the British Museum has a fine series from Grenada including 3 normals and 11 black,

[29] *Ibis*, 1912, pp. 523–8.

and also 16 blacks from St. Vincent.   If the black specimens from Grenada were put with the normals which are almost certainly nothing but a recessive form of the same bird, the variation would strike the eye on even a superficial glance at the drawer.   But following the notions so naively expressed in the passage quoted above from W. N. Lawrence, the blacks from Grenada are put apart together with the other blacks from St. Vincent, though two of them were shot on the same date as one of the normals.

CHAPTER VII

LOCAL DIFFERENTIATION. *Continued*

OVERLAPPING FORMS

The facts of the distribution of local forms on the whole are consistent with the view that these forms come into existence by the sporadic appearance of varieties in a population, rather than by transformation of the population as a whole. Of such sporadically occurring varieties there are examples in great abundance, though by the nature of the case it can be but rarely that we are able to produce evidence of a previous type being actually superseded by the variety. When the two forms are found co-existing in the same area they are usually recorded as one species if intergrades are observed, and as two species if the intergrades are absent. On the other hand when two forms are found occupying separate areas, when, that is, the process of replacement is completed in one of the areas, then forthwith each is named separately either as species or subspecies. Successive observations carried out through considerable periods of time would be necessary to establish beyond question that the history proceeds in one way rather than another. Such continuity of observation has for the most part never been attempted. The kind of information wanted has indeed only been lately recognized, and really critical collecting is a thing of only the last few decades. The methods of the older collectors, who aimed at bringing together a few typical specimens of all distinct forms, are of little service in this class of inquiry, which is better promoted by the indiscriminate collection of large numbers of common forms from many localities. When this has been done on a comprehensive scale we shall be in a position to form much more confident judgments as to the general theory of evolution.

Some little work of the kind has however been done and the results are already of great value. Seeing that the differentiation of local forms is only made possible by isolation, it necessarily happens that the collector finds one form in one locality and another in a distinct locality, and there is no evidence as to the behaviour which the two representative species might exhibit if they came into touch with each other. In the most familiar examples of such distinction each inhabits an island, completely occupying it to the exclusion of any other similar form. It can only be when the two representative species occupy parts of a continental area connected with each other by regions habitable for the organism in question, that there is a chance of seeing the two forms in contact. Often also, even where this condition is satisfied, the habits, social organisation, or some other special cause may act as a barrier which prevents the distinguishable forms from ever coming into such complete contact as to inter-breed or to behave as a genetically continuous race. When genetic continuity is ensured by a constant diffusion of the popu-lation over the whole area which they inhabit there will mani-festly be no formation of local races. The practical uniformity, for example, of so many species of birds which inhabit widely extended ranges of Western Europe is doubtless maintained by such constant diffusion. When, as in the case of the Falcons, many localities have peculiar forms, the fact may be taken as conclusive evidence that there is little or no diffusion; and when we find in such a species as the Goldfinch that in spite of mi-gratory fluctuations there are nevertheless geographical races fairly well differentiated, it may similarly be inferred that these fluctuations habitually move up and down on paths which do not intermingle. There are however a few examples of animals, not given to much irregular wandering, which occupy a wide and continuous range of diversified country and are differentiated as local races in two or more districts, though the distinct races meet in intervening areas. Of these the most notorious illus-tration which has been investigated with any thoroughness is that of the species of *Colaptes* (Woodpeckers) known in the United States as Flickers. The study of the variations of these forms, made by J. A. Allen[1] is an admirable piece of work, with which

[1] J. A. Allen, *The North American Species of the Genus Colaptes, Considered with Special Reference to the Relationships of C. auratus and C. cafer*. Bull. Am. Mus. Nat. Hist., IV, 1892.

every student of variation and evolutionary problems should make himself familiar. The two forms with which we are most concerned are known as *C. auratus* and *C. cafer*, and are very strikingly different in appearance. In size, proportions, general pattern of colouration, habits, and notes, the two are alike, but they differ in the following seven respects as stated by Allen.

Auratus	*Cafer*
1. Quills *yellow*.	1. Quills *red*.
2. Male with a *black* malar stripe.	2. Male with a *red* malar stripe.
3. Adult female with *no* malar stripe.	3. Adult female with usually a brown malar stripe.
4. *A scarlet nuchal crescent in both sexes.*	4. No nuchal crescent in either sex.
5. Throat and fore neck *brown*.	5. Throat and fore neck *grey*.
6. Whole top of head and hind neck *grey*.	6. Whole top of neck and hind neck *brown*.
7. General plumage with an *olivaceous* cast.	7. General plumage with a *rufescent* cast.

These differences are illustrated in the accompanying coloured plate, which has been most kindly prepared for me under the instructions of Dr. F. M. Chapman of the American Museum of Natural History. Before going further it is worth considering the nature of these differences a little more closely. All but the last are large differences which no one would overlook even in a hasty glance at the birds. If the only distinction lay in the colour of the quills we might feel fairly sure that *auratus* was a recessive form of *cafer*, and so probably it is in this respect. Similarly the black malar stripe of *auratus* is in all probability recessive to the red malar stripe of *cafer* and I imagine the pigments concerned are comparable with those in the Gouldian Finch (*Poephila gouldiae*) of Australia. Both sexes in that species may have the head black, red, or, less often, yellow, and though it is not any longer in question that birds may breed in either plumage, I believe that the young are always black-headed and I imagine that those which become red-headed possess a dominant factor absent from the permanently black-headed birds.[2] Yellow as a recessive

[2] For a case in which a red-headed female X a black-headed male gave a black-headed female and a red-headed male, see *Avian Mag.*, N. S., IV, pp. 49 and 329.

form of a red is certainly very common, but red and black as variants of the same pigment are less usual. In the Gouldian Finch we seem to have a case where a pigment can assume all three forms. It would be interesting to know whether the red of the malar stripes in *Colaptes* is a pigment of the same nature as the red of the quills. Both in *Colaptes* and in *Poephila gouldiae* I have seen specimens intermediate between the black and the red, and the appearance of the part affected was exactly alike in the two cases, red feathers coming up among the black ones, and many feathers containing both red and black pigments mixed together.

The development of the scarlet nuchal crescent in *auratus* and the absence of this conspicuous mark in *cafer* constitute from the physiological point of view the most remarkable pair of differences. When the red crescent is not formed, the feathers which would bear it are exactly like the rest, and no special pigment is visible in them which one can regard as ready to be modified into red. If the crescent is due to a factor it must therefore be supposed that this factor has the power of modifying the pigment of the neck in one special place alone. Dr. W. D. Miller called my attention to the fact that a similar variation occurs in another American woodpecker, the Sapsucker, *Sphyropicus varius*.[3]

I do not suggest that such variations are without parallel: indeed in *P. gouldiae* the factor which turns the black of the head into scarlet affects one special region of the black only, being sharply distinct from the unmodified black of the throat. These regions of the head are however often the seat of special colours in birds.[4] So also may be instanced the variety of the Common

---

[3] The other variations of this bird are also interesting and important. The normal male has a red head and a red throat. The female has a red head and a white throat, but varieties of the female are known with a black head, thus again illustrating the change from black to red. It should be noted that this is not a mere retention of a juvenile character, but, as the birds mature, the red feathers come up, or as an exception, the black. There is also a western species, *ruber*, in which both sexes have a great extension of red, and are alike. The male of *nuchalis* intergrades with this type, but the female does not.

[4] Dr. W. Brewster, for example, has a remarkable specimen of the Teal (*Nettion carolinense*) with a white collar strongly developed at the front and sides of the neck, in a place where the normal has no such mark.

Guillemot (*Uria troile*) which has a white line round the eyes and at the sides of the head where the normal has no such mark; but this line is formed in a very special place, the groove joining the eye to the ear, whereas the feathers of the nuchal crescent are not ostensibly distinguished from those adjacent.[5]

The transposition of the brown and the grey on the back and front of the neck also constitutes a very remarkable difference. If either grey or brown depends on a factor then it must be supposed that *auratus* has one of these factors and *cafer* the other.

From these several considerations it is quite clear that if *auratus* and *cafer* are modifications of the same type produced by presence or absence of factors, several independent elements must be concerned, and to unravel their inter-relations would be most difficult even if it were possible to breed the types under observation, which is of course quite beyond present possibilities.

The distribution of the two is as follows. On the east side of the Continent *C. auratus*, relatively pure, occupies the whole of Canada and the States from the North to Galveston. Westward it extends across the whole continent in the more northern region to Alaska, but in its pure form it only reaches down the Pacific coast to about the northern border of British Columbia. Its southern and western limit is thus roughly a line drawn from north of Vancouver, southeast to North Dakota and then south to Galveston. *C. cafer* in the comparatively pure form inhabits Mexico, Arizona, California (except Lower California and the opposite coast), central and western Nevada, Utah, Oregon, and is bounded on the east by a line drawn from the Pacific south of Washington, south and eastward through Colorado to the mouth

---

[5] This variety is spoken of as the Ringed Guillemot and is sometimes regarded as a distinct species to which the name *ringvia* was given by Brünnich. In support of this view Dr. William Brewster, to whom I am indebted for much assistance in regard to the variation of birds, called my attention to observations of his own and also of Maynard's, that the ringed birds were sometimes mated together, though in a small minority (see Brewster, *Proc. Boston Soc. N. H.*, XXII, 1883, p. 410). It would however be possible to produce many instances of varieties mated together though surrounded by a typical population (*e. g.*, two varying Blackbirds, *Zoologist*, p. 2765; two varying Nightjars, *ibid.*, p. 5278). I am inclined to believe that in nature matings between brothers and sisters are frequent in many species of animals, and that the production of sporadically varying colonies is thus greatly assisted.

of the Rio Grande or the Gulf of Mexico. Between the two lines thus roughly defined is a band of country about 1,200–1,300 miles long and 300–400 miles wide, which contains some normal birds of each type, but chiefly birds exhibiting the characters of both, mixed together in various and irregular ways. Even in the areas occupied by the pure forms occasional birds are recorded with more or less indication of characteristics of the other form, but within the area in which the two forms are conterminous, the mixed birds are in the majority. The condition of these birds of mixed character is described by Allen as follows:

"As has been long known—indeed, as shown by Baird in 1858—the 'intermediates' or 'hybrids' present ever-varying combinations of the characters of the two birds, from individuals of *C. auratus* presenting only the slightest traces of the characters of *C. cafer*, or, conversely—individuals of *C. cafer* presenting only the slighest traces of the characters of *C. auratus*— to birds in which the characters of the two are about equally blended. Thus we may have *C. auratus* with merely a few red feathers in the black malar stripe, or with the quills merely slightly flushed with orange, or *C. cafer* with either merely a few black feathers in the red malar stripe, or a few red feathers at the sides of the nape, or an incipient, barely traceable scarlet nuchal crescent. Where the blending of the characters is more strongly marked, the quills may be orange-yellow or orange-red, or of any shade between yellow and red, with the other features of the two birds about equally blended. But such examples are exceptional, an unsymmetrical blending being the rule, the two sides of the same bird being often unlike. The quills of the tail, for example, may be part red and part yellow, the number of yellow or red feathers varying in different individuals, and very often in the opposite sides of the tail in the same bird. The same irregularity occurs also, but apparently less frequently, in the quills of the wings. In such cases the quills may be mostly yellow with a few red or orange quills intermixed, or red with a similar mixture of yellow. A bird may have the general colouration of true *cafer* combined with a well-developed nuchal crescent, or nearly pure *auratus* with the red malar stripes of a *cafer*.

Sometimes the body plumage is that of *C. auratus* with the head nearly as in pure *cafer*, or exactly the reverse may occur. Or we may have the general plumage as in *cafer* with the throat and crown as in *auratus*, and the malar stripe either red or black, or mixed red and black, and so on in almost endless variations, it being rare to find, even in birds of the same nest, two individuals alike in all their features of colouration. Usually the first trace of *cafer* seen in *auratus* manifests itself as a mixture of red in the black malar stripe, either as a few red feathers, or as a tipping of the black feathers with red, or with merely the basal portion of the feathers red. Sometimes, however, there is a mixture of orange or reddish quills, while the malar stripe remains normal. In *C. cafer* the traces of *auratus* are usually shown by a tendency to an incipient nuchal crescent, represented often by merely a few red-tipped feathers on the sides of the nape; at other times by a slight mixture of black in the red malar stripe."

Such a state of things accords very imperfectly with expectations under any received theory of Evolution. As in some of the instances discussed in the first chapter we have here two fairly definite forms, nearly allied, which on any evolutionary hypothesis must have been evolved either the one from the other, or both from a third form at a time not very remote from the present, as time must be measured in evolution. Yet though intermediates exist in some quantity, no one can for a moment suggest that they are that definite intermediate from which *auratus* and *cafer* descend in common.   One cannot imagine that the immediate ancestor of these birds was a mosaic, made up of asymmetrical patches of each sort: but that is what many of the intermediates are.   It is not much easier to suppose the ancestor to have been a nondescript, with a compromise between the developed characters of each, with quills buff, malar stripes neither black nor red, with a trace of nuchal crescent, and so on. Such Frankenstein-monsters have played, a considerable part in the imaginations of evolutionary philosophers, but if it were true that there was once a population of these monsters capable of successful existence, surely they should now be found as a population occupying the neutral zone between the two modern forms. Yet, though much remains to be done in clearing up the facts, one thing is certain, namely that the neutral zone has not a definite and normally intermediate population, but on the contrary it is peopled by fragments of the two definite types and miscellaneous mongrels between them.

On the other hand, one cannot readily suppose that either form was the parent of the other. The process must have involved both addition and loss of factors, for whatever hypothesis be adopted, such changes must be supposed to have occurred. A careful statistical tabulation of the way in which the characters are distributed in the population of the mixed zone would be of great value, and till that has been done there is little that can be said with certainty as to the genetics of these characters. In the collection of Dr. Bishop of New Haven I was very kindly allowed to examine a sample, all taken at random, near together, in Saskatchewan. There were females 4 adult, 2 young; males 4 adult and 5 young. This number, though of course insufficient, is enough to give some guide as to the degree of definiteness which the characters generally show in their variations. Of the 15 birds, 8 had simply yellow quills; 2 had red; 1 was almost red but had one yellow tail-quill; 3 were intermediate and 1 was buff. As regards the malar patch, which can only be determined properly in the adult males, 1 was red, 1 was approximately red, 2 intermediate. As to nuchal crescent 4 females had none, 2 females very slight; 7 males had it, 1 had only a slight crescent, and 1 had none. In point of quills therefore 10 were definite out of 15; in point of crescent, 11 were definite out of 15; and in point of malar patch 1 only was definite out of 4. The last is a feature directly dependent on age and so counts for less, but as regards the other two features there is some indication that the factors show definiteness in their behaviour. It must be remembered that we have no knowledge what the heterozygous form may be, and in the case of red and yellow it is probably a reddish buff. The patch-works are no doubt to be compared with other well-known pied forms, and in these we must suppose the active factor broken up, which it probably can be very easily. The asymmetry, which Allen notices as so marked a feature, in the distribution of the red and yellow quills of the tail especially, recalls that of the black markings in the pied Canaries. As is well known to students of variations *some* pigment-factors in *some* animals are apparently uncontrolled by symmetry, while in other specific cases symmetry is the rule. On the other hand the blackness or redness of the malar patches is, I think, as a rule nearly symmetrical. It should be mentioned that two of Dr. Bishop's young birds belonged to the same nest, one a female with *red* quills, the other a male with *yellow*. Both are without crescent.

As to the question whether certain combinations of characters occur with special frequency, the evidence is insufficient to give a definite answer. Among all the birds I have seen in America or in England I have not yet found one having the malar patches black without any nuchal crescent. Of Dr. Bishop's 8 adults not one, however, showed the combination of the three chief features normal for *auratus* or for *cafer*.

Besides the two forms that we have hitherto considered, several other local types exist, and these throw some further light on the problem. Of these the most important in this connexion is *chrysoides*, which inhabits the whole of southern California and the mainland opposite. This remarkable form is as Allen says, very different from *auratus* except that it has the quills yellow like *auratus*, not red like *cafer*. So that we find here in the extreme west of the whole distribution a type agreeing in one of its chief features with the eastern type. Between this and *cafer* intergrades have, according to Allen, not been found. The relations of this *chrysoides* are, Allen thinks, rather with *mexicanoides*, a southern, smaller race with colours more intense, which inhabits Guatemala, but however that may be, it must be regarded as a *cafer* which has lost its red quills. The island of Guadeloupe off Lower California has an island form. Beyond the other side of the continent there is also an island form of *auratus*, inhabiting Cuba, so that clearly the yellow quills can extend into the tropics.

The above account is in many respects incomplete, but it suffices to give an outline of the chief facts. The whole problem is complicated by the undoubted effects of an uncertain amount of migration, and in many, perhaps all, districts, the winter population differs from the summer population of the same localities. The existence of these seasonal ebbs and flows is now well known to ornithologists, and most of the bird species of temperate regions are subject to them.

Difficult as it may be to conceive the actual process of origin of the two types *auratus* and *cafer*, it is I think still harder to suggest any possible circumstance which can have determined their development as distinct races, or which can maintain that distinctness when created. Some will no doubt be disposed to appeal once more to our ignorance and suggest that if we only knew more we should see that the yellow quills, the black "moustache" and the red crescent, specially qualify *auratus* for the north and eastern region, and the red quills, red "moustache" and absence of crescent fit *cafer* to the conditions of its homes. Each can judge for himself, but my own view is that this is a vain delusion, and that to cherish it merely blunts the receptivity of the mind, which if unoccupied with such fancies would be more ready to perceive the truth when at last it shall appear. Think of the range of conditions prevailing in the country occupied by *auratus*—a triangle with its apex in Florida and its base the whole Arctic region of North America. Is it seriously suggested that there is some element common to the "conditions" of such an area which demands a nuchal crescent in the Flickers, though the birds of the *cafer* area, almost equally varied, can dispense with the same character? Curiously enough, the geographical variation of *Sphyropicus varius*, another though a very different Woodpecker[6] shows that conversely the nuchal crescent can be dispensed with in the Eastern form though it is assumed by the Western.[7]

[6] The Sap-suckers feed on trees and somewhat resemble our Spotted Woodpeckers in general appearance. *Colaptes* feeds on the ground and corresponds perhaps rather with the European Green Woodpecker.

[7] For an introduction to this example I am indebted to Mr. W. D. Miller of the American Museum of Natural History. Some account of the facts is given by Baird, Brewer, and Ridgway (*A Hist. of N. Amer. Birds,* 1874, II, pp. 540, 544, etc.). *S. varius* occupies the whole country in suitable places from the Atlantic to the eastern slopes of the Rockies, and all Mexico to Guatemala. *S. nuchalis* was

Allen points out the interesting additional fact that super-posed upon each of the two distinct forms, *auratus* and *cafer*, are many geographical variations which can very naturally be regarded as climatic. Each decreases in size from the North southward, as so many species do.[8] They become paler in the arid plains, and show the ordinary phases which are seen in other birds having the same distribution. Such differences we may well suppose to be determined directly or indirectly, by environment, and we may anticipate with fuller knowledge it will be possible to distinguish variations of this nature as in the broad sense environmental, from the larger differences separating the two main types of *Colaptes*, which I surmise are altogether independent of such influences.

It is generally supposed that phenomena like those now so well established in the case of *Colaptes* are very exceptional, and as has already been stated a number of circumstances must combine in order that they may be produced. I suspect however that the examples are more numerous than is commonly thought. In all likelihood the three forms *Sphyropicus varius, nuchalis* and *ruber* are in a very similar condition though the details have not, so far as I know, been worked out. A complex example which is closely parallel to the case of *Colaptes* was described by F M. Chapman[9] at the same date as Allen's work. This is the case of *Quiscalus*, the Grackles, which in the North American Continent have three fairly distinct forms which Chapman speaks of as *Q. aeneus, Q. quiscula*, and *Q. quiscula aglaeus*. The birds are all, so far as pigment is concerned, dark blackish brown, but the head and mantle have superposed a metallic sheen of inter-ference-colours which in the various forms take different tints,

first known from the Southern Rockies only, but many were afterwards taken in Utah. *S. ruber* is restricted to the Pacific coast. In Ridgway's opinion all three are geographical forms of one species. In *ruber* the sexes are alike having both a great extension of the red in the throat, and a red crescent. The male of *nuchalis* grades to the *ruber* form, but the female does not. This female has some red in the throat like the male of *varius*, whereas the female of *varius* has a whitish throat.

[8] Not only vertebrates but the marine Crustacea and Mollusca illustrate this curious "principle" of variation, as Canon Norman formerly pointed out to me with abundant illustrations. There are of course cases to the contrary also.

[9] Chapman, F. M., *Bull. Amer. Mus.*, IV, 1892, p. 1; see also Ridgway, *Birds of North and Middle America*, 1902, Part II, p. 214.

bluish green, bronze green, or bronze purple. The details are complicated and difficult to appreciate without actual specimens, but the two common types are sufficiently distinct. The birds inhabit the whole area east of the Rockies, *quiscula aglaeus* occupying Florida and the Southern States southwest of a band of country about a hundred miles broad extending roughly from Connecticut to the mouth of the Mississippi; and *aeneus* taking the area north and west of this band. In discussing this case Chapman expresses the same view as Allen does in the *Colaptes* case, that there are two distinct populations, substantially fixed, and that the band of country in which they meet each other has a mongrel population, with no consistent type, but showing miscellaneous combinations of the character of the two chief types.

The warblers of the genus *Helminthophila* provide another illustration which has points of special interest. The two chief species are *H. pinus*, which has a yellow mantle and lower parts, white bars on the wings, a black patch behind the eyes and a broad black mark on the throat; and *H. chrysoptera* with dark grey mantle and pale whitish grey lower parts, yellow bars on the wings, and grey marks on cheeks and throat where *pinus* has black. These two birds are exceeding distinct, and in addition their songs are quite unlike. *H. pinus* ranges through the eastern United States up to Connecticut and Iowa. *H. chrysoptera* is a northern form extending down to Connecticut and New Jersey. Both are migrants.

In these two States, where the two types overlap, certain forms have been repeatedly found which have been described as two distinct species, *Lawrencei* and *leucobronchialis*. Dr. L. B. Bishop and Mr. Brewster showed me two long series of *Helminthophila* containing various intergrades between the four named kinds, and details regarding these may be found in Chapman's *North American Warblers* and in Dr. Bishop's paper in *Auk*, 1905, XXII. Though the characters evidently break up to some extent, the series can be represented as due to recombinations of definite factors more easily than the others which I have described. The differentiating characters are:

Pinus	*Chrysoptera*
1. Mantle and lower parts *yellow* (Y^1).	1. Mantle and lower parts *grey* (y^1).
2. Wing-bars *white* (y^2).	2. Wing-bars *yellow* (Y^2).
3. Cheek and throat *not black* (b).	3. Cheek and throat *black* (B).

The grey pigment of the mantle is common to both, but is masked by the yellow in *pinus*, the net result being an olive-green.[10]

I am much indebted to Dr. F. M. Chapman for the loan of the coloured plate in which these distinctions are shown. It first appeared in his book, *North American Warblers*.

We cannot tell whether *yellow* or *not-yellow* is due to the presence of a factor, but we may suppose that one or other gives the special colour to the parts. The black of character 3 is no doubt a dominant. Thus *pinus* becomes Y^1y^2b and *chrysoptera* in y^1Y^2B. The *Lawrencei* which has the underparts *yellow*, wing-bars *white*, and *black patches* is Y^1y^2B and *leucobronchialis* which has mantle and underparts *not-yellow*, wing-bars *yellow* and *no black patches* is y^1Y^2b. This representation, it should be clearly understood, is tentative and approximate only. The characters are not really sharp, for there is much grading; but allowing for the effects of heterozygosis and for some actual breaking-up of factors I believe it gives a fairly correct view of the case. In particular we can see how it meets the difficulty which Chapman felt in accepting *leucobronchialis* as in any sense derived from *pinus* which has a yellow breast, and *chrysoptera* which has a black throat, seeing that *leucobronchialis* has neither. We now recognize at once that this form could be produced by ordinary re-combination of the absence of Y^1 with the absence of B.

I note also with great interest that the modern observers agree that the so-called hybrids may have the song either of the one species, or of the other, or a song intermediate between the two. It may also be added that these two types have several

---

[10] It would aid greatly in factorial analysis if the descriptive term "green" could be avoided in application to cases where the green effect is due only to a mixture of black and yellow pigments. The absence of yellow is the sole difference between the mantle and underparts of *pinus* and *chrysoptera*.

From Chapman's Warblers of North America.

Fig. 1. *Helminthophila pinus*, **male.**

Fig. 2. *Helminthophila pinus*, **female.**

Fig. 3. "Lawrence's Warbler," male; one of the integrading forms.

Fig. 4. "Brewster's Warbler," male; another of the intergrading forms.

Fig. 5. *Helminthophila chrysoptera*, male.

Fig. 6. *Helminthophila chrysoptera*, female.

times been seen, in the breeding season, paired with each other or with one of the other combinations.

Allen[11] has described another excellent American example, the Tits of the group *Baeolophus bicolor-atricristatus*. The form *bicolor* belongs to the eastern States and ranges from the Atlantic coast to the Great Plains, and *atricristatus*, of east Mexico, extends from Vera Cruz to central Texas. In southern and central Texas the breeding ranges adjoin, and in this country various intermediates occur. The chief types differ in two main points.

B. bicolor	*B. atricristatus*
Forehead varies from deep *black* to dull black, suffused with rusty brown.	Forehead *white* to buffish white.
Crown and crest *grey*, slightly darker than the black.	Crown and crest *black*, abruptly contrasting with the back.

The intergrades between the two have, as usual, received specific names. A detailed description is given by Allen, from which it appears that the gradation is very complete. In one case a series of 16 adults were all intermediates. It is not stated whether the collector took these at random, but from the local lists it is clear that the types are found not far away from the place where the intergrades were shot.

Another very striking case is that of the Tanagers, of the genus *Rhamphocoelus*. In this group there are several local forms which are related to each other in remarkable ways. The forms known as *passerinii* and *icteronotus* exhibit the clearest phenomena of intergradation. The species *passerinii* has a brilliant scarlet and black male, and it inhabits Honduras and Nicaragua. Proceeding southwards along the isthmus we find next *costaricensis* which has a male like that of *passerinii* (but a female with more orange than the olive-grey female of *passerinii*). Next we come to Panama which is occupied by *icteronotus*, sharply distinguished from *passerinii* by the fact that the *scarlet is replaced by lemon-yellow*. This same *icteronotus* occurs again as a pure type in Ecuador and many other parts of South America; but Colombia, *between Panama and Ecuador*, contains scarlets like *passerinii*, yellows like *icteronotus*, and various intergrades

[11] *Bull. Amer. Mus. Nat. Hist.*, XXIII, 1907, p. 467.

of several shades of orange. The *passerinii* males from Nicaragua are indistinguishable from those of Colombia, and the *icteronotus* of Ecuador are the same as those in Panama. The orange intergrades, doubtless heterozygous forms, though collected at the same locality (Medellin in Colombia) as several pure yellows and pure scarlets, are in the British Museum series sorted out as a separate species under the name *chrysonotus!* Complications are introduced by the relations of these forms to another named type, *flammigerus*, but we may for our purpose leave that out of consideration, and say that the order of geographical sequence from Honduras to Ecuador is (1) scarlet, (2) yellow, (3) mixture of types, scarlet, yellow, orange, (4) yellow.

Similar examples exist in the birds of the old world, but I do not know of any that have been studied so fully as those of America. The best known is that of the two Rollers, *Coracias indicus* which spreads from Asia Minor through Persia, Baluchistan, the Indian Peninsula and Ceylon, and *affinis* which ranges from Nepal, through Assam, Tenasserim and the Indo-Chinese countries. The two types are very different and may be distinguished as follows:

C. *indicus*	C. *affinis*
Mantle drab brown-chestnut.	Dark olive-green.
Breast chestnut.	Dull purple brown.
Throat purplish, streaked with white.	Purple, streaked with blue.
Upper tail-coverts indigo.	Turquoise.

The wings are the same in both. In the provinces of Nepal, Sikhim, and Darjiling the two species coexist, with the result that intergrades have been frequently recorded. The line of intergradation extends to the coast, and birds showing various combinations of the two types from the Calcutta district exist in collections.[12] The case is interesting inasmuch as like that of *Quiscalus* it shows a series of combinations of various metallic colours. Some of these are probably evoked by the development of pigment behind striations or other interferences already existing, but in the present state of knowledge it would be quite im-

---

[12] References on this subject will be found in *Brit. Mus. Cat. Birds*, XVII, p. 13.

possible to suggest what the actual factors producing these appearances may be.

There are, naturally, many other cases among birds which are suspected of being in reality comparable, but in most of them the evidence is still inadequate. Among Lepidoptera also there are a few of these; perhaps the most striking is that of *Basilarchia* "*proserpina.*"[13] The genus is well known to European collectors under the name *Limenitis*, of which we in England have one species, *L. sibylla*, the "White Admiral." A species very like *sibylla* in general appearance is common in the northern parts of the United States, ranging through Canada and Northern New England, but rarely south of Boston. This species has the conspicuous white bands across both wings like our *sibylla*.

There is also a more Southern type known as *astyanax*, which is very different in its appearance, being without the white bands and having a broad irroration of blue scales on the posterior border of the hind wings. The two are so distinct that one would not be tempted to suspect any very close relation between them. In its distribution *astyanax* is described by Field as replacing *arthemis* south of latitude 42°. About Boston it is much more common than *arthemis*.

The two forms encroach but little on each other's territory, but where they do coexist, a third form, known as *proserpina*, is found which is almost intermediate, with the white bands much reduced. There is now no doubt that this *proserpina* is a heterozygous form, resulting from a combination of the characters of *arthemis* and *astyanax*. Field succeeded in rearing a brood of 16 from a *proserpina* mother caught wild which laid 31 eggs, and of these, nine (five males, four females) resembled the mother, being *proserpina*, and seven (four males, three females) were *arthemis*. There can be no question therefore that the mother had been fertilised by a male *arthemis* and that *no-white-band* is a factor partially dominant over the *white band*. Another point of interest which Field observed was that the *proserpina* female refused to lay on birch, poplar or willow, but accepted

[13] For these facts I am indebted to Mr. W. L. W. Field, who has lately published an account of his observations and experiments. See especially, *Psyche*, 1910, XVII, No. 3, where full references to previous publications are given.

wild cherry (*Prunus serotina*) a species on which *astyanax* can live, though that tree is not known to be eaten by *arthemis*. Incidentally also the observations show that sterility cannot be supposed to be the bar which maintains the distinctness of *arthemis* and *astyanax*.

In this connection *Papilio oregonia* and *bairdii* should be mentioned.[14] *P. oregonia* is one of the numerous forms like *machaon*, but rather paler. It is a northern insect, inhabiting British Colombia east of the Cascade Range, and reaching to Colorado. *P. bairdii* is a much darker butterfly, representing the *asterias* group of the genus *Papilio*. Like *asterias* it has the abdomen spotted at the sides, not banded as in the *machaon* group. It belongs to Arizona and Utah extending into Colorado. From Colorado the form *brucei* is described, more or less intermediate, like *bairdii* but with the abdomen banded as in *oregonia*. W. H. Edwards records the results of rearing the offspring of the *bairdii*-like and of the *oregonia*-like mothers. Each was found able to have offspring of both kinds, that is to say, *bairdii* females gave both forms, and *oregonia* females gave both forms. It is not possible to say which is dominant, since the fathers were unknown. On general grounds one may expect that the *bairdii* form will be found to dominate, but this is quite doubtful.

From this particular discussion I omit reference to those examples in which the permanently established types are obviously associated with special conditions of life. Where considerable climatic differences exist between localities, or when we pass from South to North, or from the plains into Alpine levels we often find that in correspondence with the change of climate there is a change in the characteristics of a species common to both. When I say "species" in such a connection I am obviously using the term in the inclusive sense. Some would prefer to say that in the two sets of conditions *two representative species* exist. Whichever expression be preferred it is plain that such examples present another phase of the problem we have been just considering, and in them also we have an opportunity of

[14] For the facts and further references see W. H. Edwards, *Butterflies of N. America*, 2d series, Papilio VII and X; 3d series, 1897, Papilio IV, *Can. Entom.*, 1895, XXVII, p. 239.

observing the consequences of the overlap of two closely related types, but there are advantages in considering them separately. In the examples hitherto given, with the possible exception of the Papilios,[15] the two fixed types severally range over so extensive a region that it may fairly be supposed that in the different parts they are subject to considerable diversities of climate. There is no outstanding difference that we know distinguishing the habitats of the two forms; but in comparing Alpine with Lowland forms, or essentially northern with essentially southern forms we do know an external circumstance, temperature, that may reasonably be supposed to have an influence, direct or indirect, on the population.

[15] I think this case is fairly included because the *machaon* type is so widespread that it cannot be regarded as a product of a Northern climate, nor can *asterias* be claimed as especially a warm country form, seeing that *brevicauda*, which is scarcely distinguishable from *asterias*, inhabits Newfoundland (having a curious phase there in which the yellow is largely replaced by red).

CHAPTER VIII

LOCALLY DIFFERENTIATED FORMS. *Continued*

CLIMATIC VARIETIES

In this chapter we will examine certain cases which illustrate phenomena comparable with those just considered, though as I have already indicated, they form to some extent a special group. The outstanding fact that emerges prominently from the study of the local forms is that when two definite types, nearly allied, and capable of interbreeding with production of fertile offspring, meet together in the region where their distributions overlap, though intergrades are habitually found, there is no normally or uniformly intermediate population occupying the area of intergradation. Such phenomena as these must, I think, be admitted to have great weight in any attempt to construct a theory of evolution. True we must hesitate in asserting their positive significance, but I see no escape from the conclusion that they throw grave doubt on conventional views. Again and again the same question presents itself. If A and B lately emerged from a common form why is that common form so utterly lost that it does not even maintain itself in the region of overlapping? Almost equally difficult is it, in the cases which I have numerated, to apply concrete suggestions based on any factorial scheme. We may see that in *Heliconius erato* the type with the red mark on the hind wing probably contains a dominant factor, and that where the red mark is absent the metallic colours are exposed; and that similarly the green metallic colour may have another factor which distinguishes it from the blue. In this way we can fairly easily represent the various types of *erato* on a factorial system as the result of the various possible combinations of two pairs of factors. But there we stop, and we are quite unable to suggest any reason why one area should have the red and the green type while another should have the blue also. So again with *Colaptes* or the Warblers. By application of a factorial system, admittedly in a somewhat lax fashion, the genetic interrelations of the types can be represented; but how it comes about that each type maintains a high degree of integrity in its own region we can only imagine. Each has in actual fact a stability which the intermediate forms have not, but we cannot yet analyse the nature of that stability. Mendelian conceptions show us how by segregation the integrity of the factors can be in some degree maintained, but not why certain combinations of factors should be exceptionally stable. All that is left us to fall back on is the old unsatisfying suggestions that some combinations *may* have greater viability than others, that there *may* be a tendency for like to mate with like, and so forth.

These difficulties acquire more than ordinary force in those cases in which the two fixed types inhabit regions differing in some respect so obvious and definite that we are compelled to regard each type as climatic and as specially adapted to the conditions. When for example an animal has a distinct type never met with except in Arctic or Alpine conditions, and another type proper to the plains and temperate regions, what are the characteristics of the population of intermediate latitudes or at intermediate levels? Some of the examples discussed in the last chapter may be instances of this very nature, but even if they are not, others are forthcoming which certainly are. The evidence of these cases leads to the suspicion that with further knowledge they will be found to consist of two classes, some in which the observer as he passes from the one climate to the other will find the intermediate area actually occupied by a population of intermediate character, and others in which, though we may presume the maintenance of intermediate conditions in the transitional area, there is no definite transitional population. This interrupted or discontinuous distribution seems, so far as I have means of judging, to be by far the more common of the two. I do not doubt that by sufficient search individuals representing every or almost every transitional form can be found, but it is apparently rare that *populations* corresponding to these several grades can be seen. The question has in few if any cases been studied with precision sufficient to provide a positive answer;

but I suspect that real and complete continuity, in the sense thus defined, will only be found where the character of the local populations depends *directly* on the conditions of life, and shows an immediate response to changes in them apart from that postponed response which we suppose to be achieved by selection. Obviously the character must be one, like size for instance, capable of sensibly complete gradation.

The only example I have met with of the phenomenon of anything like a complete intergradation between local types really distinct in kind is that provided by the butterfly *Pararge egeria*. It is well known to entomologists that this insect exists in two very different types, a northern one, the "Speckled Wood" of England, in which the spots are a pale whitish yellow, and a southern type having the full fulvous colour that we know as characteristic of *megaera*, the "Gatekeeper." It appears that Linnaeus gave the name *egeria* to the southern type,[1] and our own is now called *egerides*. Broadly speaking, so far as Great Britain, France, and the Spanish Peninsula are concerned, the tawny-coloured *egeria* occupies Spain and western France up to the latitude of Poitiers and the pale yellow *egerides* extends from Scotland, where it has a scanty distribution, through southern England, where in suitable localities it is common, and the north of France to Paris.[2] The two types when placed side by side are strikingly different from each other, and are an excellent illustration of what is meant by climatic variation. The insect is not a great traveller and probably scarcely ever wanders far from its home. It should therefore be possible by collecting from north to south to find out how the transition is effected, whether suddenly or gradually. This at various times I have endeavoured to do, but I am still without exact information as to the population in certain critical areas. In addition to the information derived from specimens which I have collected or seen in the collections of others there is a good account of the general distribution in Europe given by the Speyers,[3] who evi-

[1] Often referred to by older writers as *Meone*, Esper's name.

[2] There are also two distinct island forms, unlike the European, *Xiphia* of Madeira, and a smaller variety, *Xiphioides* of Canary. See especially, Baker, G. T., *Trans. Ent. Soc. London*, 1891, p. 292.

[3] Speyer, Adolf, and August. *Verbreitung der Schmetterlinge*, 1858, I, p. 217.

dently paid more attention to the subject than most lepidopterists have done, and many more recent records. In particular Oberthür[4] has published many details as to the distribution in western France and I am especially indebted to Mr. H. Rowland-Brown for a long series of notes as to the distribution in France generally, and to Mr. H. E. Page and Dr. T. A. Chapman, Mr. Oberthür Prof. Arrigoni degli Oddi, Mr. H. Williams and other correspondents, for showing me forms from many localities. The butterfly is attached for the most part to woods of deciduous trees and to country abounding in tall hedges or rough scrub. It is not usually to be found in highly cultivated districts or in very dry regions. Hence there is necessarily some want of continuity in the distribution at the present time and I should think a mile or two of arable land without big hedges would constitute a barrier hardly ever passed. The larva feeds on several coarse grasses, especially *Dactylis glomerata*. Barrett mentions also *Triticum repens*. In this country the winter is usually passed in the larval stage, but I have found that in captivity, at least, there is much irregularity. The larvæ feed whenever the weather is not very cold and may pupate, but if sharp cold comes on when they are pupating or nearly full-grown they often get killed unless protected.

Some writers speak of a difference between the early and later broods, but I have never noticed this, and I do not think that the general tone of the yellow is affected by the seasons (see Tutt, *Ent. Rec.*, IX, 1897, p. 37).[5]

Beginning at the south of Spain the thoroughly fulvous type *egeria* is common at Gibraltar in the Cork woods, at Granada, and doubtless generally. Lederer is said to have found only this type in Spain (Speyer), and though I have no precise information as to other places in the Peninsula north of Jaen I feel tolerably sure that there is no change from south to north.[6]

[4] *Lepid. Comparée*, fsc. III, p. 372.

[5] Mr. Rowland-Brown has called my attention to a statement by Dr. Vaillantin (*Petites Nouv. Ent.*, II, 235) that in Indre-et-Cher the first brood is of the northern type and the second of the southern. My experience is that in captivity these distinctions do not occur, and I have true *egeria* as first brood from Vienne and as the late brood from the Landes. I never collected in Indre-et-Cher.

[6] I have since seen true *egeria* from Ferrol in the extreme northwest, which was in Mr. Tutt's collection.

Immediately north of the Pyrenees we still meet *egeria* exclusively, and up to Poitiers at least there is no noticeable change. But somewhere between Poitiers and the bottom of the Loire valley at Tours, the genuine southern type comes to an end, and the whole population begins at the Loire to be of an intermediate type, easy to distinguish both from *egeria* and from *egerides*. As to the exact condition of the species in the fifty miles separating St. Savin on the Vienne from places on the Loire I have no adequate information. I have only one small sample from there, but it does contain insects both of the southern and intermediate types taken on the same day, in a wood near Preuilly. Oberthür also states that at Nantes the true southern form exists in company with the northern. From this I infer that the southern form extends up the coast further than it does inland, but I imagine the representative spoken of as northern would be of usual Brittany or intermediate type.

The Vienne river joins the Loire, so the true southern type reaches over into the basin of the Loire. From the Loire (Tours, Cormèry) north to Calvados (Balleroy) only the intermediate is found, so far as I know, and the same type extends over Brittany.[7] In general, however, the woods near Paris have the thoroughly northern type *egerides*, but at St. Germain-en-Laye and at Etampes (Oberthür) the population approaches the intermediate type.

On the whole the intermediate type is certainly less homogeneous than either of the extremes, and females with the two central spots either paler or more fulvous than the rest are not uncommon, but I have never taken one on the Loire or in Brittany which I should class with either of the extreme types.

Before speaking of the distribution in other parts of France and in Europe generally I will briefly state the results of my breeding experiments. The work was done many years ago before we had the Mendelian clue, and it is greatly to be hoped that some one will find opportunities of repeating it. Crossing the English and the thoroughly southern type the families produced

---

[7] Mr. G. Wheeler kindly showed me a series identical with this type, from Guernsey, and others from near Laon.

agree entirely with the intermediates of Brittany and the Loire. Reciprocals are alike. Of F_2 I only succeeded in raising very few and of those that I had (about 30) nearly all were intermediate in character, though perhaps rather less uniform than F_1. One family alone, containing only 4 specimens, had one *egerides*, and three fulvous intermediates. As the case stands alone I hesitate whether or not to suppose it due to some mistake. Moreover from F_1 crossed back with the respective parental types I had fairly long series, especially from $F_1 \times$ the southern type, and looking at these families I cannot see any clear evidence of segregation. On the contrary, I think that though there are slight irregularities, they would, taken as a whole, be classed as coming between the intermediate type and the extreme form used as the second parent. This at least is true when the second parent was of the southern type.

On this evidence I have regarded the case as one in which there is no good evidence of segregation and as conforming most nearly with the conventional view of gradual transition in response to climatic influences. Such influence must however be indirect; for I reared five generations of the northern type in England, and these, though they included several abnormal-looking specimens in the last generation and then died out, did not show any noticeable change from the fulvous colour of the wild type. Merrifield[8] also found that heat applied to pupae of the northern type produced no approach to the southern type.

Looking at the facts now in the light of more experience it seems to me just possible that the case may be one in which, as in Nilson-Ehle's Wheats, the dominant differs from the recessive in having two pairs of factors with similar effects. The fulvous type for example may have two or more elements in separate pairs which together produce the full effect, and the intermediate may have one of these. If this were so, some segregation should of course eventually be observable, but the proportion of the various fulvous and fulvous-intermediate individuals would be large, and the reappearance of actual repre-

[8] *Ent. Rec.*, V, 1894, p. 134.

sentatives of the northern type might be rare. I admit that this is a somewhat strained interpretation of the facts, and as yet it is not entitled to serious consideration. Nevertheless I am led to form some such expectation partly from the great difficulty in the way of any other, partly from the evidence of the small mixed sample found at Preuilly and partly from the statements given by Oberthür. There are moreover other features in the general distribution of the species which make it improbable that the dependence on climate can after all be so close. Published lists are unfortunately of little use in deciding which form occurs at a particular place, because, since the name *Meone* has ceased to be used for the southern form, there is no complete unanimity among authors as to the application of the names *egeria* and *egerides*, and unless more particulars are given, either name may be used for either form. Besides this, difficulty arises from the fact that the intermediate type is not generally distinguished at all, and English collectors finding it, may easily record it as the southern type. From Staudinger's note on the distribution, I gather that he, on the contrary, reckoned the intermediate with the northern type, as do the Speyers also. The late Mr. J. W. Tutt was careful to distinguish the three forms and has left several useful records. Easy therefore as it might seem to be to make out the distribution of such a familiar insect in its various modifications, there are serious practical difficulties, and until long series are brought together with this special object in view many obscurities will remain.

With only the series from England, the west of France, and Spain before one it would be easy to regard the successive series of tones as a fair measure of climate; the brighter the colour, the hotter might one expect the locality to be. Such rough correspondence is often to be observed in butterflies and birds. It becomes impossible to take these simple views in the light of more complete knowledge. Beginning with France the fulvous *egeria* occupies the lower valley of the Rhone, probably from well above Lyon, though I have no exact information respecting the country above Avignon. According to Speyer it also takes the department of Lozère. The same authority says that Puy-de-

Dôme has "*egeria*," meaning perhaps the intermediate form, with the fulvous form much less commonly. Next comes the curious fact that though the Lower Rhone (Avignon, Tarascon, Nîmes) has the true fulvous form, Hyères, Cannes, Grasse, Nice, Digne, and Alassio have *the intermediate.* Savoy has the intermediate (Chambéry) and even *egerides* perhaps, though in the same latitude on the west of France there is nothing but the fulvous type. At Chalseul and Besançon (Doubs) the ordinary northern type is found. Switzerland generally, I believe, has the northern type, but Staudinger gives *egeria* for Valais and the intermediate occurs in Vaud.[9] The south side of the Alps has probably colonies of the pale *egerides;* and of intermediates. Orta, with a very hot summer, has the English type (Tutt, *Ent. Rec.*, XII, 1900, p. 328). Locarno has the intermediate (*ibid.*, XV, 1903, p. 321). North Italy in general and western Piedmont have the intermediate; but further south *egeria* begins, at what region I do not know. Speyer gives on his own authority the remarkable statement that at Florence both extremes occur, but chiefly intermediates between the two. Mr. R. Verity however kindly informs me that in his experience this is not so, and that neither the real southern type nor the northern occur there. Sardinia, Sicily, Crete all have the southern type. Greece probably has various types. Staudinger (*Hor. Ross.*, VII, 1870, p. 78) says intermediates resembling Nice types common everywhere, but from "Greece" the British Museum has a series that would pass for English specimens; and the same type occurs near Constantinople. The island of Corfu has a pale intermediate, distinct from *egerides* but approaching it. In Roumania all three forms are recorded from various places: *egeria* in the Dobrutscha; not quite typical (presumably an intermediate) at Bukharest; intermediate in various mountainous localities as well as in Macedonia and Dalmatia; but *egerides* in Azuga at about 3,000 feet.[10] Hungary has the true *egerides* also. (Cf. Caradja, *Deut. Ent. Zt.*, IX, p. 58.) Mathew records the same

[9] Mr. Wheeler has some pale but rather worn specimens from the Rhone Valley at Vernayaz.

[10] See Fleck, E., Die Macrolep. Rumäniens, *Bul. Soc. Sciinte*, VIII, 1899, p. 720.

from Gallipoli (*E. M. M.*, 1881, p. 95). Staudinger does not distinguish the intermediates from the northern, but he gives "*egerides*" for Armenia and Fergana (Central Asia). As against the mere proximity of a great mountain chain being the influence which keeps the Riviera population intermediate may be mentioned the fact that the northern foothills of the Pyrenees have the pure southern type, and the climate of Cambo must surely be far cooler than that of Nice. The exact locality of the Greek specimens is not given, but there can be no part of Greece which is not much hotter in summer than Brittany, or Calvados, which have the intermediate, not the English type.

In face of these facts it can scarcely be maintained that average temperature is the efficient cause of the particular tone of colour which the butterfly shows in a given region. Nevertheless it is clear that climate counts for much in determining the distribution. It is noticeable that though the pale *egerides* can be established in a warm climate we never find *egeria* in cold climates, and even the intermediate is not found in places that have a hard winter. I suspect that the distribution of the broods through the year and the condition of the animal at the onset of hard frost are features which really determine whether a strain can live in a particular place or not. Though the truth of the suggestion cannot be tested by experiments in captivity, which at once introduce disturbances, I incline to the idea that *egeria* has not got the right periodicity for northern climates. If it could arrange its life so that the population consisted either of young larvae, or perhaps of thoroughly formed pupae[11] at the onset of winter, it might, for any obvious reason to the contrary, be able to live in England. It is irregularly "polyvoltine," as the silk-worm breeders say, and as soon as a little warmth encourages it, a new generation starts into being, which if the frost comes at an untimely moment, is immediately

[11] My experience agrees with that of Mr. H. Williams (*Ent. Rec.*, VIII, 1896, p. 181) that pupae, well-formed, can stand considerable frost; but I used to find that half-grown larvae usually died if unprotected, and I believe that larvae which attempted to pupate in warm autumn weather and then got caught by frosts, always died. Small larvae which can creep into shelter at the bottom of the plants survived, and I expect that in the north the winter is usually passed in that state (see also Merrifield, F., *Ent. Rec.*, VIII, 1896, p. 168, and Carpenter, J. H., *ibid.*).

destroyed. Many species are continually throwing off individuals which feed up fast[12] and emerge at once if the temperature permits, and I imagine a species of Satyrid wholly or largely represented by such individuals could scarcely survive in a country which had a hard winter. For such a climate some definite periodicity in the appearance of the broods may well be indispensable. But assuming that *egeria* is cut off from cold climates for such a reason, there is nothing yet to connect these habits with the fulvous colour, and until breeding can be carried out on a satisfactory scale there is no more to be said.

From time to time records appear of individual specimens more or less fulvous being caught in southern England, especially in the New Forest.[13] It would be interesting to know what offspring such individuals might produce. From the evidence now given some notion both of the strength and the weakness of the case considered as one of continuous climatic variation can be formed. I know no other equally satisfactory. Whether or not definite mixture of the intermediates with either of the extremes will be proved to occur, the case differs materially from those considered in the last chapter in the fact that at all events there is no general overlapping of forms. In a species so little given to wandering, overlapping could indeed scarcely be expected to occur. It is this circumstance which makes the species preeminently suitable as a subject for the study of climatic influences, and I trust that entomologists with the right opportunities may be disposed to explore the facts further.

Just as many species, like *egeria*, have varieties which can be regarded as adapted to northern and southern regions, so there are also several which have lowland and Alpine forms quite distinct from each other. Every such case presents an example of the problem we have been considering. As the collector passes from the plains to the Alpine region, how will he find the

---

[12] Some most unlikely species do this. I once had a larva of *Parnassius delius*, found at about 5,500 feet, which emerged late in the autumn (in October I believe), a season at which it must have perished in its own country.

[13] See, for examples, Barrett, G. C., *Lepidoptera of the Brit. Islands*, I, 1893, p 229; also Grover, W., *Ent. Rec.*, IX, 1897, p. 314; Williams, H., *Proc. Ent. Soc.*, 1898, who reared several specimens from the New Forest which would pass for Bretons, though the rest of the family were true *egerides*.

transition from one form to the other effected? Does the low-land form give place to the Alpine form suddenly, with a region in which the two are mixed, or will he find a zone inhabited by an intermediate population? I have spent a good deal of time examining the facts in the case of *Pieris napi* and its Alpine female variety *bryoniae*, and though there are many compli-cations which still have to be cleared up, no doubt is possible as to the main lines of the answer. If in any valley in the Alps inhabited by both *napi* and *bryoniae* the collector catches every specimen he can, beginning at the bottom and working up to 7,000 feet, he will at first get nothing but *napi*. At about 2,500 feet, he may catch an occasional *bryoniae* flying with the *napi*. After 3,000 feet *napi* usually ceases, and only *bryoniae* are found. As an exception a colony of *napi* may be met with at much greater heights. I once found them in numbers at about 6,000 feet.[14] Not only were they free from any trace of modification in the direction of *bryoniae*, but they were of the thoroughly southern type of *napi*, being a late brood of that large and very pale kind (*meridionalis*) almost destitute both of dark veining above and of green veining below, which are common on the shores of Lago Maggiore and in other hot southern localities. Not far off at the same level were typical *bryoniae* in fair abund-ance. Occasionally an intermediate may be met with. I have taken a few, for example, at Macugnaga and at Fobello. These, however, in my experience are rarities in the Alps. Fleck[15] gives notes on the distribution in Roumania which shows the same state of things. The lowland form is not transformed though found at great heights, and at Azuga (nearly 3,000 feet) *bryoniae* occurs with only occasional "*flavescens*," viz., inter-mediates of the second brood.

If this were all the evidence we should be satisfied that the lowland and Alpine types keep practically distinct, overlapping occasionally, but rarely interbreeding. The problem would remain, how is the distinctness of the two types maintained in the region of overlapping? Nowadays, I suppose, we should

[14] Above the Tosa falls.
[15] *Bul. Soc. Sciinte*, VIII, 1899, p. 691.

incline to answer this question by reference to segregation, and perhaps by an appeal to selective mating. The suggestion that segregation does take place is certainly true to some extent. There are, however, difficulties in the way, and the whole subject is one of great complexity. My own experiments were made in pre-Mendelian times and were not arranged with the simplicity which we now know to be essential. The results are neither extensive enough nor clear enough to settle the many collateral questions which have to be considered, and the work ought to be done again. Nevertheless, some notes of the observations may have a suggestive value.

When I began, I did not sufficiently appreciate that the "*napi*" group, omitting the North American forms, and the Asiatic representatives, has at least three chief types in western Europe. The differences we have to deal with are manifested by the females only, so in this account particulars as to the males are omitted for the most part. These are (1) our own British *napi;* (2) the form found in the south, from the Loire downwards, and in the Italian Alps, which I think may be spoken of as *meridionalis;* (3) *bryoniae*, which is a form clearly recognizable in the *female* only, and is found only in the arctic regions and in the Alps above 2,500 feet. The first two have several broods, two, three, or more, according to opportunity, and the first brood is different from the later ones. In *napi* the markings on the upper surface are a dark grey but in *meridionalis* they are a pale silvery grey and much less extensive. In the later broods of *napi* there is much less general irroration of the veins, and the spots stand out as more defined and blacker. These differences vary greatly in degree of emphasis. In *meridionalis* the later broods are entirely different from the first. Instead of having silvery markings they have the ground colour quite white, with the spots large and a full black. On the under side of the hind wings the usual green veins are almost absent, and I have seen individuals which could scarcely be distinguished from *rapae*. To these later broods the term *napaeae* is sometimes applied, but I here use *meridionalis* for the southern race in general as applicable to all broods.

The female *bryoniae* is totally unlike the others. The ground colour is a full yellow, and each nervure is thickly irrorated with a brown pigment often spreading so far as to hide the ground almost entirely in the forewings. The males corresponding with these females are not certainly distinguishable from those of our own *napi*. Both sexes have the green veining of the underside of the hind wing fully developed, rather more than is usual in the lowland races, but this is not really diagnostic of the variety. The first serious difficulty arises in regard to the second brood of *bryoniae*. It is stated that there is only one brood,[16] but I feel fairly sure that a second brood is sometimes produced, and that the females with a yellow ground and diminished irroration of the veins, not very uncommon in the Italian Alps in July to August, are generally representatives of it. Such insects would of course be classed with *bryoniae* in collections.

My experiments began with eggs of true *bryoniae* females caught at about 2,500 feet early in July. These emerged in August–September as intermediates with yellow ground and about half as much black on the upper surface as *bryoniae*. They are exactly like the intermediates usually found in nature and in the light of later experience I regard them as natural F_1 forms, and I think the mothers had been fertilised by *napi* males, though I admit that in view of the rarity of natural intermediates there is a difficulty in this suggestion. Three of these females were mated with males raised from thorough *meridionalis* females, and three families were produced. Two of them showed distinct evidence of segregation, some being yellow and some white with various intergrades, some being no blacker than *meridionalis* and some ranging up to a dark intermediate type. Part emerged in the same autumn; and part overwintered, emerging as the spring *meridionalis* or as the peculiar type which I afterwards learnt to know as the spring F_1 form. The distinctions were fairly sharp between the several forms. But the offspring of the third female gave a series practically continuous from

[16] The fact that Weismann by heating pupæ obtained only one autumn specimen seems to me to show rather that a second brood can be produced than that it cannot, which is the inference usually drawn.

meridionalis to the F_1 type. The work of subsequent years gave results similarly irregular which could only be described adequately at great length. The outcome may however be summed up in the statement that there is evidence that both the yellow ground and the dark veining are due to factors, but that there are several of these and that imperfect segregation is not uncommon, producing various reduction-stages. The yellow ground may be due to one factor, and the several shades may be the result of irregularities in dominance, but the black markings when fully developed cannot I think be the result of less than three factors, one for the basal darkening, one for general irroration, and one for the margins. Probably also the enlargement of the spots is produced by a fourth factor.

There was not, in my experience any great difficulty in getting the various forms to pair in captivity. Some attempts were made to see whether individuals of either type selected mates of their own type in preference to those of the other, but the results were inconclusive. There were some indications of such a preference; though, from the impossibility of judging how much of this may be due to other circumstances, I could not come to a positive conclusion on the rather meagre evidence.

Recently Schima[17] has given a careful and detailed account of all the forms found in Lower Austria which he enumerates under 14 distinct varietal names. He gives full references to previous accounts, especially to the beautiful plates lately published by Roger Verity.[18] Examination of these and of my own specimens strongly suggests that the several forms are due to the recombination of the factors I have named. Among those which I have bred are representatives of most if not all the types enumerated by Schima in addition to other curious forms. For example I have *bryoniae* markings on a ground practically white; the dark veins with spots almost obsolete; *meridionalis* on a yellow ground; the intermediate amount of black on a white ground, etc. The last-named may occur wild and I have one from Macugnaga as well as one given me by Mr. F. Gayner from Lulea (Lapmark).

[17] Schima, K., *Verh. Zool. bot. Ges. Wien.* LX, 1910, p. 268.
[18] *Rhopalocera Palaearctica*, Florence, 1905-11, especially Pl. XXXII.

13

To obtain really exact knowledge of the number of factors and their properties it would be necessary to repeat the work. After the beginning, I made a mistake in using British *napi* instead of *meridionalis* and the results were much confused thereby. The contrast between *meridionalis* and the various dark forms is much greater and classification of the types would have been therefore easier. The British form is presumably *meridionalis* plus the factor for the basal pigmentation. The problem is greatly complicated by the differentiation of the seasonal forms. The first point to be determined is whether *bryoniae* is capable of producing a second brood when it is thoroughly pure-bred, and whether such a second brood is, as I suspect, normally intermediate in character.

In the Alps generally there is no definitely intermediate population; nor I believe, is any such population met with in the north where the arctic *bryoniae* meets *napi*, but as to this I have no precise information. One curious fact, however, must be mentioned, namely that there is a population that can probably be so described with fairness established at Mödling near Vienna. This is not in any sense an Alpine locality, and does not, as I am told, differ in any obvious way from the other suburbs of Vienna. Dr. H. Przibram was so good as to send me a set taken at this place, representing a second brood, and they were decidedly heterogeneous, ranging from an intermediate form such as *bryoniae* fertilised by *napi* usually produces, to a light yellowish second-brood type with little dark pigment. There are also two actual *bryoniae*. Whether true *napi* also occur there I do not know, but I have no doubt they do. It would be well worth while to investigate the Mödling population statistically, and to breed from the intermediates which might not impossibly prove to be heterozygotes. There are also records of such intermediates being occasionally found in some parts of Ireland, in the north of Scotland, and in south Wales,[19] but I do not know of any regular colony of these forms. We can scarcely avoid the inference that one or more of the factors which make up *bryoniae* may be carried by these intermediates. It is not

[19] See figures in Barrett, G. C., *Lepidoptera of Brit. Islands*, I, pt. 3, p. 25.

clear why their interbreeding does not produce actual *bryoniae* occasionally. If this occurred, the probability is that the fact would be known to collectors, at least in the British localities. The absence of true *bryoniae* must, I think, be taken to mean that some essential factor is absent from these intermediates.

To sum up the evidence, the facts that are clear may be thus enumerated:

1. *Napi* and *bryoniae*, or in the Italian Alps, *napaeae* and *bryoniae* frequently meet each other.

2. They cross without difficulty, producing fertile offspring.

3. But in the levels at which they overlap there is no intermediate population, and only occasional intermediate individuals.

4. In certain parts of the distribution of *napi* similar intermediates sometimes occur, and at one place (Mödling) they are so frequent as apparently to constitute a colony.

5. As to the genetic relations of the two forms there is no complete certainty. Indications of segregation have been observed in some cases, but there are several factors concerned and they are liable to some disintegration.

Another form in which I tried to investigate the same problem is *Coenonympha arcania*, which has one Alpine form known as *darwiniana*, and another, *satyrion*. In calling *satyrion* a form of *arcania* I follow Staudinger and other authorities, but I have never been quite satisfied that it should be so regarded. The differences between *arcania* and *darwiniana* are essentially differences of degree; *C. arcania* occurs in places where there is cover, and reaches up the valleys usually as high as the mixed woods of deciduous trees, which is about 2,500 feet. The variety *darwiniana*, on the contrary, is an insect of treeless hillsides, and I regard it as a dwarf and possibly a stunted form. It would not greatly surprise me to find that with the application of good conditions *arcania* could be raised from *darwiniana* eggs, or that if *arcania* larvae were starved they might give rise to *darwiniana* butterflies. I have been unsuccessful in trying to rear the species, having lost the larvae by disease. Usually one does not catch *arcania* and *darwiniana* on the same ground, and as *Festuca ovina* —a typically hill-side grass—is a common food-plant of *dar-*

winiana there can be little doubt that *arcania* feeds on some other grass, probably woodland species. Colonies of *arcania* of varying size and brightness are commonly found, and though a sample of *arcania*, finely grown, from a warm Italian wood, presents a striking contrast with *darwiniana* from an Alpine pasture, one certainly may get samples which fill all the gradations. Generally the sample from a given locality is fairly homogeneous.

Of *satyrion* I have little personal experience. I only twice found it, namely at Zinal, and at Hallstatt in Austria, but it occurs at Zermatt, Arolla, and in several Swiss localities above 5,000 feet, and I understand that it is the typical Alpine form in the Engadine. With its darkened colour and reduced size it might well be expected to be a still further stunted form of *darwiniana*. Yet I have never found the one succeed to the other at the higher levels. If *darwiniana* appears when Alpine conditions are reached in a valley it will be met with up to the highest level at which such butterflies live. Tutt was of opinion that *satyrion* is a distinct species.[20] I once, at the top of the Vorderrheinthal caught a sample of *darwiniana* a few of which (males) were so dark and had the eye spots so poorly developed that they looked like transitions to *satyrion*. Otherwise I never found any such transitional forms and they are certainly exceptional. There is further a record[21] of *satyrion* having been taken flying with *arcania*. This was near Susa, at about 2,000 feet I infer. Mr. H. E. Page has similar specimens from Caud and from St. Anton (Arlberg). The females, however, both of mine and of Mr. Page's samples are a pale brown, quite unlike the females both of *arcania* and of the dark Zinal *satyrion*. The difficulty thus raised has not I think yet been considered by the authorities, and it is possible that the Alpine forms of *arcania* are in reality three, not two.

The evidence taken together suggests, I think, that *darwiniana* is related to *arcania* much as so many of the Alpine varieties

[20] Tutt, J. W., *Ent. Rec.*, XVIII, 1905, p. 5. In the same place he states that on the Mendel Pass *arcania* "runs into" *darwiniana* and that in the Tyrolean localities the transition is especially evident. Wheeler (*ibid.*, XIII, 1901, p. 121) expresses the contrary opinion, that *satyrion* does grade to *arcania*.

[21] H. Rowland-Brown, *Ent. Rec.*, XI, 1899, p. 293.

of plants are to the well-developed individuals of the lower levels. I do not anticipate that factorial differences will be found in these insects, and it is by no means impossible that the distinctions between them are the direct consequences of altered conditions. The relations of *arcania* to *satyrion* are more doubtful, and in that case a factorial difference may at least be suspected.

The species of the genus *Setina* have Alpine forms which agree in possessing a characteristic extension of the black pigment to form radiating junctions between the spots on the wings. Speyer, who discussed the interrelations of these forms in detail,[22] lays stress on the absence of genuine transitional forms between *aurita* and the variety *ramosa*. Both are mountain insects but *ramosa* extends to levels higher than that at which *aurita* ceases, which is about 4,000 feet. The two forms are often found flying together. Speyer says that his brother searched diligently for transitional forms at the level of overlapping, but found none, so that at least they may be regarded as rare. The variety *ramosa* is not infrequent at much lower levels (*e. g.*, Chiavenna, 1,020 feet; Reussthal, 1,500 feet) and extends as high as the permanent snows. In the British Museum collection, however, I have seen several that I should regard as transitional. Speyer perhaps would have classed as *ramosa* all in which the spots of the central field were united, and it is by no means unlikely that breeding would prove such individuals to be heterozygous.[23]

[22] Speyer, Stettiner, *Ent. Ztg.*, XXXI, 1870, p. 63.

[23] In regard to the closely analogous case of *Spilosoma lubricipeda*, Standfuss makes a similar statement. He bred the type on a large scale with the radiate form which he calls *intermedia*, and says that in four years of miscellaneous crossing he never obtained really transitional forms. Nevertheless after examining large series, especially those of Mr. W. H. B. Fletcher, I came to the conclusion that several might be so classed, but I am quite prepared to find that such specimens are heterozygous. (See Standfuss, *Handb. d. Gross-Schmet.*, 1896, p. 307.) It is by no means unlikely that various dark forms of *lubricipeda* correspond with a progressive series of factorial additions. Many of the stages have been named, and of these the most definite are the *intermedia* of Standfuss (probably = *eboraci* of Tugwell) and the very dark *Zatima* of Heligoland, in which only the thorax, the nervures and a small field in the forewings remain yellow. A form was bred by Deschange from *Zatima* in which even the field in the forewing is obliterated. The exact circumstances in which *Zatima* occurs in Heligoland would be worthy of special investigation, for the

There can scarcely be a doubt that the distinction between *aurita* and *ramosa* is factorial, the radiate *ramosa* probably having the factor for striping. In support of this view may be mentioned the observation of Boisduval,[24] respecting a gynandromorphous individual, which was *aurita* male on one side, and *ramosa* female on the other. Speyer makes another excellent comment. He points out that the simple notion that the radiation is a mere extension of pigmentation consequent on the climate of the higher levels, will not fit the facts very easily, because the size of the spots varies greatly in *aurita* itself at any level, and lowland specimens may actually have more black confined to the spots alone than some *ramosa* possess on spots and lines combined.[25]

The two Salamanders, *S. maculosa* and its Alpine form *atra*, might not improbably furnish evidence bearing on the same problem. The two are of course very distinct, not merely in colour (*maculosa* being spotted with yellow or orange while *atra* is entirely black) but also in the mode of reproduction, a feature to which reference will be made in the next chapter. I cannot, however, find any evidence as to the overlapping of the two forms. *S. atra* occurs from about 3,000 feet or somewhat less, and reaches great elevations in the Eastern Alps, but I do not know if the two forms ever occur in the same localities. Leydig,[26] Boulenger,[27] and most modern authorities regard the two types as distinct species, but they are in any case closely allied, and it would be of interest to have exact knowledge of their geographical delimitations.

The reader who has considered the cases adduced will appreciate the difficulties which must be faced in any attempt to

normal *lubricipeda* in also found on the island. For references as to the British occurrences see especially, Hewett, W., *Naturalist*, 1894, p. 353. As to *Zatima* see especially Krancher, *Soc. Ent.*, II, 1887–8, p. 26. I am indebted to Dr. Hartlaub for information as to the Heligoland types.

[24] Boisduval, *Bull. Soc. Ent. Fr.*, III, 1834, p. 5.

[25] The systematics of *Setina* have been much controverted, but no one I believe doubts that *aurita* and *ramosa* are forms of one species. See also Chapman, A. T., *Ent. Rec.*, XIII, 1901, p. 139.

[26] *Arch. Naturg.*, 33, 1867, p. 116.

[27] *Brit. Mus. Cat., Batrachia Gradientia*, 1882.

account for the facts in a rational way. As always in a problem of Evolution, two separate questions have to be answered. First how did the form under consideration come into existence, and secondly, how did it succeed in maintaining itself so as to become a race? The evidence from the local forms, though very far from giving complete answers to either of these questions definitely refutes the popular notion that a new race comes into existence by transformation of an older race. If a gradual mass-transformation of this kind took place we should certainly expect that when two types, nearly allied and capable of interbreeding, overlap each other in their geographical distribution, a normally intermediate population would exist. If each type can maintain itself and if each came into existence by gradual transformation, then there must have been an intermediate capable of existing and maintaining itself as a population; and if this had ever been, surely in the region of overlapping, that intermediate population should continue. Especially should such a population be found when the two extreme types are adaptational forms and the region of overlap is a region of intermediate conditions. But of the examples we have examined there is only one, that of *Pararge egeria* and *egerides*, which can at all be so interpreted, and even in that case it is not impossible that more minute observation would reveal discontinuity between the extremes and the admittedly normal intermediate population. Granting provisionally however that this example, as it stands, is consistent with the conventional theory of evolution, I know not where we should look for another case equally good. When the distinctions are produced by direct influence of conditions operating during the lifetime of the individuals, examples of intermediate populations occupying the areas of intermediate conditions can no doubt be produced. Many turf-like Alpine plants, for instance, if protected from exposure and properly nourished can grow as large as those of the same species found in the valleys, and in the case of such quantitative effects, intermediate conditions can doubtless produce intermediate characters.

Even these examples however are not very abundant, and often the intermediate locality has not a form intermediate

between those of the two extreme localities, but some third
form distinct from either. This is the case for instance in the
fauna of brackish waters. We are taught to believe that the
fresh water fauna was evolved from the marine fauna, which
it well may have been; but as students of Crustacea and Mollusca
know familiarly, the brackish water forms are not as a rule intermediates between fresh water species and sea species, but more
usually they are special forms belonging to the brackish waters,
with the peculiar property that they can tolerate a great range of
conditions, and live without ostensible variation in waters of
most various compositions and densities, which very few marine
or fresh water species are able to do.

Sometimes the distinction between local races, as in *Rhamphocoelus passerinii* and *icteronotus* may be regarded with confidence as due to one simple Mendelian factor possessed by one
race and absent from the other, but I think, more often, as in
Colaptes or in the varieties of *Pieris napi*, the existence of several
distinct factors is to be inferred. As we have seen, the races
of *Colaptes* show almost beyond doubt that in different areas at
least three distinct factorial combinations can be perpetuated
as races.

In the distribution of variability we find, I think, some hint
as to the steps by which the phenomena under consideration
have come to their present stage, and I am disposed to regard
the facts so well attested in the case of our own melanic moths
as a true indication of the process. Following this indication
we should regard the change in the character of a population
as beginning sporadically, by the appearance of varying individuals, possibly only one varying individual, in, it may be, one
place only. As to *why* a variety should increase in numbers we
have nothing but mere speculation to offer, and for the present
we must simply recognise the fact that it may. That such survival and replacement may reasonably be taken as an indication
that the replacing race has some superior power of holding its
own I am quite disposed to admit. Nevertheless it seems in
the highest degree unlikely that the outward and perceptible
character or characters which we recognise as differentiating the

race should be the actual features which contribute effectively to that result.

In discussions of geographical distribution in relation to problems of origin it is generally said that very nearly allied species usually occupy distinct areas, while other competent observers state the exact contrary. Lately, for example, Dr. R. G. Leavitt[28] has published an important collection of evidence upholding the latter proposition, taken chiefly from the botanical side, showing how in numerous genera two or more closely allied species coexist, frequently without intermediates, in the same localities, and may even be thus found in company throughout their distribution. The difference of opinion evidently arises from a confusion as to the sense in which the term "species" is understood and applied. Leavitt, for example, is avowedly following Jordan and, among moderns, Sargent, in applying a close analysis, and denoting as species all forms which are distinct and breed true. Against this use of the term I know no valid objection[29] but it must be obvious that if others follow a different practice confusion may result when observations are summarised in general statements. We will consider this subject again in another place, but here it may be sufficient to say that there can scarcely now be a doubt that numbers of these associated species, such as Jordan discriminated, represent various combinations of the presence and absence of Mendelian factors. This does not in any way weaken the argument which Leavitt founds upon the facts, namely, that the observed distribution of these forms is consistent with the supposition of an evolution largely discontinuous.

On the other hand, those who have come to the opinion that nearly allied species generally occupy distinct ground are presumably more impressed by the characters differentiating the geographically distinct or adaptational races, seeing that genuine intermediates between them are less commonly found. Those geographical races may no doubt contain various differentiated forms; but when all live together, occasional intermediates are

[28] The Geographical Distribution of nearly related Species. *Amer. Nat.*, XLI. 1907, p. 207.
[29] See later, p. 242.

usually to be found even in the case of characters habitually segregating. These segregating forms Jordan would certainly have determined as species, and it must be conceded that no physiological definition has yet been drawn which consistently excludes them.

CHAPTER IX

THE EFFECTS OF CHANGED CONDITIONS: ADAPTATION

In the attempt to conceive a process by which Evolution may have come about, the first phenomenon to be recognized and accounted for is specific difference. With that recognition the outline of the problem is defined. The second prerogative fact is adaptation. Forms of life are *on the whole* divided into species, and these species *on the whole* are adapted and fit the places in which they live. To many students of Evolution, adaptation has proved so much more interesting and impressive than specific diversity that they have preferred it to the first place in their considerations.

Whether this is, as I believe, an inversion of the logical order or not, there is one most serious practical objection to such preference, that whereas specific diversity is a subject which can be investigated both by the study of variation and by the analytical apparatus which modern genetic science has developed, we have no very effectual means of directly attacking the problems of Adaptation.

The absence of any definite progress in genetics in the last century was in great measure due to the exclusive prominence given to the problem of Adaptation. Almost all debates on heredity centered in that part of the subject. No one disputes that the adaptation of organisms to their surroundings is one of the great problems of nature, but it is not the primary problem of descent. Moreover, until the normal and undisturbed course of descent under uniform conditions is ascertained with some exactness, it is useless to attempt a survey of the consequences of external interference; nor as a rule can it be even possible to decide with much confidence whether such interferences have or have not definite consequences. Those, for example, who debated with enthusiasm whether acquired characters are or are not transmitted were constantly engaged in discussing occurrences which we now know to be ordinary features of descent under uniform conditions, and the origin of variations which were certainly not caused directly by circumstances at all. In the absence of any factorial analysis, or of any conception of what factorial composition means and implies, no one knew what varieties might be expected from given parents. The appearance of any recessive variety was claimed as a consequence of some treatment which might have been applied to the parents. There was no possible standard of evidence or means of controlling it, and thus the discussion was singularly unfruitful. Before we can tell how the course of descent has departed from the normal, we must know what the normal would have been if we had let alone. We are still far from having such knowledge in adequate measure, but it does now exist in some degree, and we are steadily approaching a position from which we shall be able to form fairly sound estimates of the true significance of evidence for or against the proposition that environmental treatment can produce positive disturbances in the physiological course of descent.

Thus described, the field for consideration is very wide. Though the effects of changed conditions were especially studied in the hope of solving the problem of adaptation by direct observation, that, as all are now agreed, is but a part of a more general question. We must ask not only do changed conditions produce an *adaptative* response on the part of the offspring, but whether they produce any response on the part of the offspring at all. It is not in doubt that by violent means, such as starvation or poisoning of the reproductive cells, effects of a kind, stunting and deformity for instance, can be made evident, just as similar effects may follow similar treatment during embryonic or larval life. Apart from interferences of this class, are there any that may be reasonably invoked as modifying the course of inheritance?

No epitome of the older evidence for the inheritance of adaptative changes is here required. That has often been collected, especially by Weismann, who exposed its weaknesses so thoroughly as to carry conviction to most minds, and showed that whether the phenomenon occurs or not, no one can yet prove that it does. Belief in these transmissions, after being almost universally held, was with singular unanimity abandoned. This change in opinion, though doing credit to the faith of the scientific community in evidential reasoning, is the more remarkable inasmuch as the strength of the idea was not derived from the minute amounts of supposed facts now demolished. On the contrary, it was really an instinctive deduction from a wide superficial acquaintance with the properties of animals and plants. They *can* accommodate themselves to circumstances. They *do* make responses sometimes marvellously appropriate to demands for which they can scarcely have been prepared. What more natural than to suppose that the permanent adaptations have been achieved by inherited summation of such responses? No one had actually been driven to believe in the inheritance of adaptative changes because bitches which had been docked had been known to give birth to tailless puppies, or because certain wheat in Norway was alleged to have become acclimatized in a few generations. Evidence of this kind was collected and produced rather as an ornamental appendix to a proposition already accepted, and held to be plainly demonstrated by the facts of nature. Looked at indeed in that preliminary and uncritical way, the case is simply overwhelming. Those who desire to see how strong it is should turn to Samuel Butler's *Life and Habit,* and even if in reading they reiterate to themselves that no experimental evidence exists in support of the propositions advanced, the misgiving that none the less they may be true is likely to remain. Making every deduction for the fact that the wonders of adaptation have been grossly exaggerated, and that marvels of fitness and correspondence between means and ends have grown out of mere anthropomorphic speculations, there is much more left to be accounted for than can at all comfortably be accepted as the product of happy accidents. So oppressive are these difficulties that we can scarcely blame those who imagine that the study of heredity is primarily directed to the problem of the transmission of acquired characters, a preconception still almost universal among the laity.

But since the belief in transmission of acquired adaptations arose from preconception rather than from evidence, it is worth observing that, rightly considered, the probability should surely be the other way. For the adaptations relate to every variety of exigency. To supply themselves with food, to find it, to seize and digest it, to protect themselves from predatory enemies whether by offence or defence, to counter-balance the changes of temperature, or pressure, to provide for mechanical strains, to obtain immunity from poison and from invading organisms, to bring the sexual elements into contact, to ensure the distribution of the type; all these and many more are accomplished by organisms in a thousand most diverse and alternative methods. Those are the things that are hard to imagine as produced by any concatenation of natural events; but the suggestions that organisms had had from the beginning innate in them a power of modifying themselves, their organs and their instincts so as to meet these multifarious requirements does not materially differ from the more overt appeals to supernatural intervention.

The conception, originally introduced by Hering and independently by S. Butler, that adaptation is a consequence or product of accumulated *memory* was of late revived by Semon and has been received with some approval, especially by F. Darwin. I see nothing fantastic in the notion that memory may be unconsciously preserved with the same continuity that the protoplasmic basis of life possesses. That idea, though purely speculative and, as yet, incapable of proof or disproof contains nothing which our experience of matter or of life at all refutes. On the contrary, we probably do well to retain the suggestion as a clue that may some day be of service. But if adaptation is to be the product of these accumulated experiences, *they must in some way be translated into terms of physiological and structural change*, a process frankly inconceivable.

To attempt any representation of heredity as a product of memory is, moreover, to substitute the more obscure for the less. Both are now inscrutable; but while we may not unreasonably aspire to analyse heredity into simpler components by ordinary methods of research, the case of memory is altogether different. Memory is a mystery as deep as any that even psychology can propound. Philosophers might perhaps encourage themselves to attack the problem of the nature of memory by reflecting that after all the process may in some of its aspects be comparable with that of inheritance, but the student of genetics, as long as he can keep in close touch with a profitable basis of material fact, will scarcely be tempted to look for inspiration in psychical analogies.

For a summary of the recent evidence I may refer the reader to Semon's paper[1] where he will find a collection of these observations described from the standpoint of a convinced believer. At the outset one cannot help being struck by the fact that of the instances alleged, very few, even if authentic, show the transmission of acquired modifications which can in any sense be regarded as adaptative, and many are examples not so much of a transmission of characters produced in the parents as of variation induced in the offspring as a consequence of treatment to which the parents were submitted, the parents themselves remaining apparently unmodified. No one questions the great importance of evidence of this latter class as touching the problem of the causes of variation, but it is not obvious why it is introduced in support of the thesis that acquired characters are inherited.

It is most difficult to form a clear judgment of the value of the evidence as a whole. To doubt the validity of testimony put forward by reputable authors is to incur a charge of obstinacy or caprice; nevertheless in matters of this kind, where the alleged phenomena are, if genuine, of such exceptional significance, belief should only be extended to evidence after every possible source of doubt has been excluded. We believe such things when we must, but not before. At the very least we are entitled to require that confirmatory evidence should be forthcoming from independent witnesses. So far as I have seen, this requirement is satisfied in scarcely any of the examples that have been lately published, and until it is, judgment may reasonably be suspended.

In some cases, however, the facts are not doubtful. Standfuss, by subjecting pupae of *Vanessa urticae* to cold, produced

[1] Semon, R., Der Stand der Frage nach der Vererbung erworbener Eigenschaften, published in *Fortschr. der naturw. Forschung.*, Bd. 11, 1910.

the now well-known temperature-aberrations in which the dark pigment is greatly extended. He put together in a breeding-cage 32 males and 10 females showing this modification in various degrees. Two of these females died without leaving young. Seven produced exclusively normal offspring. From the eighth female 43 butterflies were bred, and of these there were four (all males) which to a greater or less extent exhibited the aberrational form.[2] The mother of this family was the most abnormal of the 10 females originally put in.

Fischer's experiment with *Aretia caja* was on similar lines. From pupae which had been frozen almost all the moths which emerged showed aberrational markings. A pair of these mated and produced 173 young which pupated. Those which emerged early were all normal, but of those which emerged late, 17 had in various degrees abnormal markings like those of the parents.[3] In neither of these examples is there any question as to the facts. Both observers have great experience and give full details of their work.

As regards *Vanessa urticae*, however, it must be recalled that Fischer himself showed that in Nymphalids somewhat similar aberrations could be produced both by heat and by cold, and even by centrifuging the pupae. Frl. von Linden produced a transitional form of the same aberration in *V. urticae* by the action of carbonic acid gas.[4] It is highly probable that the appearance is due to a morbid change, perhaps an arrest of development, which may be brought about by a great diversity of causes. In the experiments the cause probably was a diseased condition of the tissues of the mother herself. She had been subjected to freezing sufficiently severe to prevent the proper development of the pigments and some of the ovarian cells presumably suffered also. It will be observed that the only specimens which were affected were the offspring of the most abnormal female, and of them only four out of forty-three showed any change.

The same interpretation probably applies to the cases in

[2] Standfuss, M., *Denks. Schweiz. naturf. Ges.*, XXXVI, 1898, p. 32.

[3] Fischer, E., *Allg. Ztschr. f. Entomologie*, Bd. VI, 1901.

[4] Out of 12 pupae treated 8 died and of the 4 survivors, one only was affected. See M. v. Linden, *Archiv. Rassen. u. Gesells.*, 1904, I.

Arctia caja. In this species the markings are well known to be liable to great variation. As Barrett says, even in nature individuals are rarely quite alike, and an immense number of strange forms occur in collections.[5] These are greatly sought after by some collectors, especially in England, where they fetch high prices at auctions, and it is notorious that most of them come from Lancashire and the West Riding of Yorkshire. It is commonly supposed that the breeders of that district subject them to abnormal conditions, and especially to unnatural feeding, but I know no clear evidence that this is true. From whatever cause it is certain that the natural pattern is, in some strains at all events, very easily disturbed.

The elaborate experiments of Schröder with *Abraxas grossulariata* are difficult to follow and are complicated by the fact that the series which was submitted to abnormal temperatures was derived from an abnormal original pair. From the evidence given it is not clear to me whether the temperature had a distinct effect. This insect, like *Arctia caja*, produces an immense number of variations (especially in the amount of the black pigment) and as most of these are, I believe, reared in domestication for sale, it is highly probable that the species is easily influenced by cultural conditions.

Schröder describes two other experiments which have been accepted by Semon and other supporters of the view that acquired characters are transmitted. In the first, *Phratora vitellinae*, a phytophagous beetle living on the undersides of leaves, was used. It naturally feeds on *Salix fragilis*, a species without a felt, or tomentum, on the underside of the leaves. Larvae were transferred to another willow (near *S. viminalis*) which has the undersides of the leaves felted. The larvae took readily to the new food, pushing the tomentum before them as they gnawed the leaves. They came to maturity and when they were about to lay their eggs they were given a free choice between *S. fragilis* and the tomentose species. The greater number of ovipositions,

[5] For illustrations see *Oberthur's Études d'Entom.*, 1896, where many of these curious aberrations are represented; also Barrett, *Lepid. Brit. Islands*, II, pp. 71 and 72.

219, took place on *fragilis*, and there were 127 on the tomentose bush, which we are told was six times as large as the *fragilis*. The larvae from *fragilis* were next put on the tomentose species and reared on it. When they became imagines they were similarly given their choice, with the result that there were 104 ovipositions on the tomentose species and only 83 on *fragilis*. In the next generations there were 48 ovipositions on the tomentose and 11 on *fragilis*. Finally the fourth generation made 15 ovipositions on the tomentose and none on *fragilis*.

The difficulty about such experiments is obviously that one has no assurance that the change of instinct, in so far as there is any, may not be a mere consequence of the captivity. It must, besides, be extremely difficult to arrange the experiment so that there is really an equal choice between the two bushes, when one stands beside the other. Przibram, in quoting this case, considers that as the tomentose bush was about six times as large as the *fragilis*, some indication of the relative attractiveness of the two may be obtained by dividing the ovipositions on the larger bush by six, but I imagine the matter must be much more complex.

Schröder's second example is not more convincing, in my opinion, though Semon regards it as one of the most important pieces of evidence. It concerns a leaf-rolling moth, *Gracilaria stigmatella*, the larva of which is said normally to make its house by bending over the *tips* of the sallow leaves on which it feeds. Schröder placed larvae on leaves from which the tips had been cut, and these larvae made their houses by rolling over the *sides* of the leaves. Their offspring were again fed on leaves without tips, and as before, they rolled in the leaf-margins either on one side or both. The offspring of this second generation were then fed on entire leaves. There were 19 houses made by these (?19) larvae, and of them 15 were normal, made by folding down the tips of the leaves, while 4 were abnormal, made by rolling in the leaf-margins. Schröder says that in nature he has only twice seen abnormal houses; but it is clearly essential not only that the frequency of such variability in nature should be thoroughly examined, but also that we should know whether when the species is bred in captivity these irregularities of behaviour do or do not occur when the larvae are fed on uninjured leaves.

The famous case of Schübeler's wheat is revived by Semon. The story will be familiar to most readers of the literature of the subject. Briefly it is that annuals, especially wheat and maize, raised from seed in Central Europe take more time in coming to maturity and ripening than similar plants raised in Norway, where the summer days are much longer. The received account is that he imported seed especially of maize and of wheat from Central Europe to Norway and found that in successive years the period of growth and ripening was increasingly reduced. After two generations seed of the accelerated wheat was sent back to Breslau where it was grown, and was found to ripen rather more slowly than in Norway, but much more quickly than the original stock had done. The facts recorded by Schübeler[6] are that he received seed from Eldena, which is on the Baltic near Greifswald. The variety is described as " *100 tägiger Sommer Weizen*," but no more exact record of its behaviour in Germany is given. This wheat, grown at Christiania in 1857, took 103 days to harvest. Its seed was again grown in Christiania in 1858, and took 93 days, and sown again in 1859 it took only 75 days, 28 days less than in the first year of cultivation in Norway. Seed of the 1858 crop was sent to Breslau, and grown there by Roedelius in 1859; it took 80 days. Evidently before such a record can be used as proving an inheritance of acquired characters numbers of particulars should be forthcoming. The view that Johannsen has taken is that the result was probably due to unconscious selection of the earlier individuals among a population consisting of many types of various compositions. Some effect may no doubt be ascribed to that cause, but I cannot think that alone it would account for the results. My impression is rather that they were produced by differences in the cultivation and especially in the seasons. Research of an elaborate character would be necessary in order to eliminate the various sources of error, and nothing of the kind has been done; nor does Semon allude to these difficulties in prominently adducing Schübeler's evidence. A

⁶ Schübeler, F. C., *Die Culturpflanzen Norwegens*, 1862, especially pp. 24 and 28.

difference of even three weeks in time of harvesting may easily be due to variation in the season. It would in any case be difficult to analyse the meteorological conditions, and to decide how much effect in postponing or accelerating the harvest might be due to cold days, to cloudy days, to wet weather, to fluctuations in average temperature, to hot days, and other such incidents occurring at the different periods of growth, even if they were specially watched while the experiments were in progress, and at this distance of time such analysis is practically impossible. Without careful simultaneous control-experiments this evidence is almost worthless. The director of the Meteorological Office[7] has, however, kindly sent me some details of the weather at Breslau from 1857 to 1860, and I notice that as a matter of fact July, 1859, was an exceptionally hot month, *having an average of 2.67° C. above the mean* for the twenty years 1848–1867. June in that year was slightly (0.31° C.) below the mean and May slightly above it (0.18° C.). August was also abnormally hot, 2.35° C. above the average. The Breslau wheat was sown on *May 19* and harvested on August 6. There was a cold spell from May 11 to 14, which this wheat escaped, as it was sown on May 19. In the other years the cold spell came much later. These elements of the weather may possibly have done something to hurry the ripening in 1859. It is unfortunate that we are not told how long similar wheat from Breslau seed took to ripen in that year.

As regards the Norway cultivations we have the average monthly temperatures recorded by Schübeler, though he does not discuss them in connection with this special problem. It is quite clear that 1857, in which the period was 103 days, was an exceptionally cold summer, especially as regards the months of June and July, but though there was, so far as the temperature

[7] I am obliged to him and to Dr. E. Gold for much trouble taken to answer my questions. Some idea of the kind of weather indicated by an average of 2.7°6 C. above the mean may be got from a comparison with the year 1911, which most people will remember as one of the hottest summers they have known. The July of that year was in east and southeast England about 4° F. above the mean but 2.67 C. means about 4.8° F. above the mean. At Greenwich July, 1859, was about 6.5° F. above the average.

records go, no great difference between 1858 and 1859, the year 1859, in which the period of ripening was the shortest, was somewhat colder in Norway than 1858. But we have the further difficulty that there were ten days difference in sowing, for in 1858 the sowing was made on May 14, and in 1859 on May 24. With all these possibilities uncontrolled, and indeed unconsidered, I am surprised that Semon should claim these experiments as one of the chief supports for his views.

Schübeler's other allegations respecting the influence of climate on plants grown in various places and especially at different elevations in Norway have been destructively criticised by Wille[8] to whose paper readers interested in the subject should refer.

Before the appearance of Wille's criticisms Wettstein[9] made a favourable reference to Schübeler's work, accepting his conclusion. He states also that he has himself made analogous experiments with flax, finding that the length of the period of development and a series of morphological characters show an adaptation to local conditions, and that on transference of seed to other conditions the previous effects are maintained. No details, however, are given, and I do not know if anything more on the subject has appeared since. The other examples cited by Wettstein, such as the observations of Cieslar on forest-trees and those of Jakowatz on gentians seem to me open to all the usual objections applicable to evidence of this kind. Such work, to be of any value for the purpose to which it is applied, must be preceded by a study of the normal heredity and of the variations of the species.

Most of the recent writers (Semon, Przibram, etc.) on the inheritance of acquired characters accept the story of Brown-Séquard's guinea pigs, which are said to have inherited a liability to peculiar epileptiform attacks induced in their parents by various nervous lesions.

The question has been often debated and several observers have repeated the experiments with varying results, some failing

---

[8] Wille, N., *Biol. Cblt.*, XXV, 1905, p. 521.

[9] Wettstein, R. von, *Der Neo-marckismus u. seine Beziehungen zum Darwinismus*, Jena, 1903.

to confirm Brown-Séquard, others finding evidence which in various degrees supported his conclusions. Recently a new and especially valuable paper has been published by Mr. T. Graham Brown[10] which goes far towards settling this outstanding question. He states that "the Brown-Séquard phenomenon is nothing more or less than a specific instance of the scratch-reflex," and it is due to a raised excitability of the mechanism of this reflex. This raised excitability is the character acquired as a consequence, for instance, of the removal of part of one great sciatic nerve. The nature of this raised excitability and its causation are discussed and elucidated, but this part of the work is not essential to the present consideration. Mr. Graham Brown in his summary of conclusions remarks that it is very difficult to see how this condition of raised excitability can be transmitted to the offspring, and this comment which might be made in reference to any of the alleged cases certainly applies with special cogency to the present example.

He then calls special attention to three observations:

1. That guinea pigs which had a "trophic" change in the foot, as a result of division of the great sciatic nerve, have repeatedly been seen to nibble the feet of other guinea pigs which had this change in the foot from the same causes.

2. That accidental injury to the toes may be followed by the Brown-Séquard phenomenon in an otherwise normal animal.

3. That in several instances the young of guinea pigs which exhibited the phenomenon have been noticed to have one or more toes eaten off by the mother.

Brown-Séquard noticed that almost all his animals in which the great sciatic was divided acquired the "epilepsy" and nibbled those parts of their feet in which sensation had been lost. Of the offspring of such animals he found that a very small proportion exhibited a malformation of the feet, and of these some showed the "epilepsy." The proportion which showed the "epilepsy" was one to two per cent. of the offspring.

Morgan[11] is quoted by Graham Brown as having suggested

---

[10] T. Graham Brown, *Proc. Roy. Soc.*, 1912, vol. 84, *B*, p. 555. This paper gives full reference to the previous literature of the subject.

[11] Morgan, T. H., *Evolution and Adaptation*, New York, 1903.

that the loss of toes in the offspring may have been due to mutilation by the mother, following his experience in a case in which the tails of mice in succeeding litters were thus devoured, and there can be little doubt that in this suggestion lies the clue to the explanation of the whole mystery. Graham Brown concludes that it may be supposed with every degree of probability that the " transmission " was due to injuries inflicted upon the young by their parents. With this conclusion most people will now be disposed to agree, and we may hope that we shall hear the last of this curious myth — to the elucidation of which a vast quantity of research has been devoted.

The series of experiments made by Kammerer with various Amphibia have attracted much attention and have been acclaimed by Semon and other believers in the transmission of acquired characters as giving proof of the truth of their views. With respect to these observations the chief comment to be made is that they are as yet unconfirmed. Many of the results that are described, it is scarcely necessary to say, will strike most readers as very improbable; but coming from a man of Dr. Kammerer's wide experience, and accepted as they are by Dr. Przibram, under whose auspices the work was done in the Biologische Vesuchsanstalt at Vienna, the published accounts are worthy of the most respectful attention.

The evidence relates chiefly to three distinct groups of occurrences:

1. Modification in *Alytes obstetricans*, the Midwife Toad, affecting both the structure and the mode of reproduction, induced by compulsory change of habits.

2. Modification in the mode of reproduction of *Salamandra atra* and *maculosa* induced by compulsory change of habits.

3. Modification in the colour of *Salamandra maculosa* induced by change in the colour of the soil on which the animals were kept.

1. I will take first the case of *Alytes*,[12] because it is the most

---

[12] Kammerer's chief paper on this subject is in *Arch. f. Entwm.*, 1909, XXVIII, p. 447, and it is to this that the paginal references in the present text relate. His previous paper appeared, *ibid.*, 1906, XXII, p. 48. An account of his further experiments with *Alytes* is given in *Natur*, 1909–10, Heft 6, p. 95.

definite example, and because it is the case which most readily admits of repetition and verification.

The habits of *Alytes obstetricans* are well known. The animals copulate on land. As the strings of eggs leave the female they are entangled by the hind legs of the male, and being adhesive they stick to him and undergo their development attached to his back and legs. The number of eggs varies from 18 to 86, a number much smaller than is usual in toads and frogs which lay their eggs in water. The eggs are large and full of yolk.

There are two breeding seasons, one about April and the other about September, and a winter hibernation. Not only animals brought in from outside, but their offspring reared in domestication maintain these normal habits in confinement, if the temperature does not exceed 17° C. (pp. 499 and 534).

If, however, the temperature be artificially raised and kept at 25–30° C., the males do not attach the eggs to themselves when spawning occurs on land but let them lie. The adhesion of the eggs is said to be hindered by the comparatively rapid drying of their surfaces.

More usually in the high temperatures the animals *take to the water* and copulate there. The eggs are ejected into the water, and as their gelatinous coverings immediately swell up, they do not stick to the males.

The offspring thus derived from the parents subjected to heat for one breeding-period only, whether they were laid in water or on land, did not show departures from the normal type.

Kammerer states next, however, that in subsequent breeding-periods the same parents frequently take to the water to breed, though they have become quite accustomed to the heated chamber; and furthermore that if such animals, having thus lost their instinct to brood their young, be transferred to ordinary temperatures they do not readily reassume their normal habits, but for several breeding seasons — at least four — will take to the water. These parents lay from 90 to 115 eggs, which are small and contain little yolk, and the larvae, on hatching, breathe with their embryonic gills until they are absorbed instead of being broken off as normally.

The offspring thus abnormally developed when they mature are said never to brood their eggs. If they are derived from the earlier spawnings of their parents, before, that is to say, the parents had been submitted to the changed conditions long enough to transmit their effects, they lay on land; but if they are derived from the later spawnings, they lay in the water. These changes of habit are manifested without the continued application of the abnormal experimental conditions, and, as I understand the account, in normal conditions of temperature.

If the abnormal experimental conditions are continued, the toads always lay in water, and their eggs become progressively smaller and more numerous. The larvae in the fourth generation acquire three pairs of gills instead of one pair, and are in other respects also different from the normal form.

Respecting the *Alytes* bred in this way Kammerer makes the very striking statement *that the males in the third generation* (p. 535) *have roughened swellings on their thumbs and that in the fourth generation* (pp. 516 and 535) *these swellings develop black pigment.* Together with the appearance of this secondary sexual character there is hypertrophy of the muscles of the fore-arm. To my mind this is the critical observation. If it can be substantiated it would go far towards proving Kammerer's case. *Alytes*, among toads and frogs, is peculiar in that the males do not develop these lumps in the breeding season, and the fact may no doubt be taken to be correlated with the breeding habits, copulation occurring on land and not in water as is usual with Batrachians. It is to be expressly noticed that these lumps on the thumbs or arms of male toads and frogs are not merely pigmented swellings, but are pads bearing numerous minute horny black spines, which are used in holding the females in the water. The figures which Kammerer gives (Taf. XVI, figs. 26 and 26*a*) are quite inadequate, and as they merely indicate a dark patch on the thumbs it is not possible to form any opinion as to the nature of the structure they represent.

The systematists who have made a special sudy of Batrachia appear to be agreed that *Alytes* in nature does not have these structures; and when individuals possessing them can be produced for inspection it will, I think be time to examine the evidence for the inheritance of acquired characters more seriously. I wrote to Dr. Kammerer in July, 1910, asking him for the loan of such a specimen[13] and on visiting the Biologische Versuchsanstalt in September of the same year I made the same request, but hitherto none has been produced. In matters of this kind much generally depends on interpretations made at the time of observation; here, however, is an example which could readily be attested by preserved material. I notice with some surprise that in a later publication on the same subject no reference to the development of these structures is made (see below).

The statements here given represent but a small part of Kammerer's papers on the subject. He gives much further information as to the course of the experiments, especially in regard to the fate of the eggs laid on land and the aberrations induced in them by treatment. The ramifications of the experiments are, however, very difficult to follow, and as I am not sure that I have always understood them I must refer the reader to the original.

More recently Kammerer has published[14] a most curious account of experiments in crossing his modified and abnormal *Alytes*, derived from the water-eggs, with normal individuals.

In the first case the cross was made between a *normal female* and an *abnormal male*. The offspring were normal in their habits. In the next generation bred from these almost exactly a quarter showed the abnormal instinct.

The reciprocal cross was made between an *abnormal female* and a *normal male*. In this case the offspring were abnormal in their behaviour; but the second generation bred from them showed three quarters abnormal and one quarter normal.

Certain details as to numbers and sexes of the various families bred in the course of this amazing experiment are given in a

[13] In reply to my letter Dr. Kammerer who was then away from home very kindly replied that he was not quite sure whether he had killed specimens of *Alytes* with "*Brunftschwielen*" or whether he only had living males of the fourth generation, but that he would send illustrative material.

[14] Kammerer, P., *Natur*, 12 December, 1909, Heft 6, p. 95, repeated in *12 Flugschrift d. Deutsch Ges. f. Züchtungskunde*, Berlin, 1910.

subsequent publication.[15] This later paper goes somewhat fully into the question of the difference in behaviour between the normal and modified individuals, describing the ways in which the males and females possessing the acquired character could be recognised from the males and females which were normal, but in this account I find no reference to the development of the "*Brunftschwielen*" — the horny pads on the hands of the males. As these structures would be of special value in such a diagnosis the omission of any allusion to them calls for explanation. Kammerer claims the evidence as proof of Mendelian segregation in regard to an acquired character, the first example recorded. Pending a repetition of the experiments there is no more to be said.

2. *The Mode of Reproduction of Salamandra atra and maculosa.*[16] — *Salamandra maculosa*, the common lowland form, with yellow bands or spots, deposits its young in water, generally as gill-bearing tadpoles, with a wide, swimming tail, though occasionally they are born still enclosed in the egg-capsule out of which they soon hatch. Spawning extends over a considerable period, often many weeks, and during the season one female may bear more than 50 young.

S. atra, the black Alpine form, produces its young on land. They are born without gills, ready to breathe air, and with the rounded tail of the adult. These differences may, as Kammerer says, naturally be regarded as adaptations to the Alpine conditions. Moreover, the female bears *only two* young in a season, and this reduction in the number must be taken to be a consequence or condition of viviparity. There are many eggs in the ovary, but all except the two which are destined to develop degenerate and form a yolk-material on which these two survivors feed.

Kammerer gives a long account of the various conditions to which he subjected both species. The treatment was complicated

[15] *Festschrift zum Andenken an Gregor Mendel*, being vol. XLIX of the *Verh. Naturf. Ver. in Brünn*, 1911, p. 98.

[16] Kammerer's chief papers on this subject are *Archiv für Entwm.*, XVII, 1904, and *ibid.*, XXV, 1907. An epitome of results is also given by him in *12 Flugschrift d. Deutsch. Ges. f. Züchtungskunde*, Berlin, 1910.

in many ways, but the essential statements are, as regards *S. maculosa*, that when no water was provided in which the young might be born, they were dropped on land, larger and in a later stage of development and of a darker colour than is normal; that the larvae so born gradually diminished in number until only two were deposited in each breeding-period; that dissection showed that the other ova degenerated to form a yolk-material. The larvae so produced reached maturity. The summary of results describes their behaviour, stating that they produced:

(*a*) *In water*, either (1) *very* advanced, large-headed larvae 45 mm. long (instead of 25–30 mm.) with gills already reduced, which had awkward, embryo-like movements, and in some few days metamorphosed into small perfect salamanders; or (2) moderately advanced, properly proportioned larvae, 40–41 mm. long, provided with large gills of (at first) intrauterine character, which were reduced during aquatic life.

(*b*) *On land*, small (26 mm. long) larvae with rudimentary gills, having the body rounded instead of being flattened from above downwards, and an elongated narrow head, which were unable to live in deep water. These larvae changed to the salamander colour in 10–12 days, and after four weeks metamorphosed into salamanders 29 mm. long.

(*c*) In the foregoing cases the experimental conditions were not continued, or in other words, basins of water were provided in which they could spawn. But if the experimental conditions are continued, these *Salamandra maculosa* which were born newt-like (viz., not in a larval condition), are themselves newt-bearing from the first time they give birth, using the dry land, and bringing forth only two young, the normal number for the births of *S. atra*. These young are 40–41 mm. long, and are dark-coloured, resembling greatly the normal new-born *S. atra*.

This epitome of the observations illustrating the inheritance of acquired characters has been very widely quoted, and may not unnaturally be taken to summarize a wide experience of the modified animals. Reference to the details given in the same paper shows that, as alleged, each of the four types of behaviour enumerated was witnessed *once* only in the case of each of four females, no two agreeing with each other. As to the number of the males or their habits nothing is said. The first female, a (1), bore five young; the second, a (2), bore two, of which one was a partial albino; the third, b, produced four young; and the fourth, c, two as already stated.

In the case of c the details show that the female gave birth immediately after being transferred from the open-air terrarium to one indoors, which contained no basin of water. This is the example of the consequences which follow on a continuance of the experimental conditions.[17]

As regards S. atra the converse is reported. Various means were used to induce them to eject their young prematurely in water, such as massaging the sides of the mothers, or raising the temperature to 25° or 30° C., with various degrees of success. But afterwards it was found that specimens collected wild at an elevation of about 1,000 metres responded to much simpler treatment, and gave birth prematurely in water when they were kept in a large shallow basin of water not so deep but that they could everywhere touch the bottom with their feet and keep their heads above the surface. With specimens collected at higher elevations this treatment was inoperative, and the suggestion is made that S. atra at the lower confines of its habitat partakes more of the nature of maculosa than do the individuals from greater heights; for Kammerer argues that pools suitable for breeding must be more uncommon at those elevations than they are lower down.

In the earlier paper[18] Kammerer states that newly caught females of S. atra often give birth in the water, and show an undoubted preference for doing so. He describes also how he once saw several females, wild in their natural habitat, lay their young in a rain-puddle at 1,800 metres elevation, but the larvae thus born were fully formed.

When the deposition of the young as larvae has become

[17] "Bei Fortdauer der Versuchsbedingungen sind als Vollmolche geborene Salamandra maculosa gleich bei der ersten Geburt abermals voll molchgebärend, benutzen zum Geburtsakt das trockene Land, und zwar unter Erreichung der (bei Salamandra atra normalen) Embryonen-Zweizahl," Kammerer, 1907, p. 49.

[18] 1904, p. 56.

" habitual "[19] with *S. atra*, three to nine larvae may be produced at one spawning period, from 35 to 45 mm. long, with gills at most 8 mm. long, and a tail-fin 2–3 mm. broad. Such larvae are generally coffee-brown, or grey (instead of black), and show other minor differences.

The summary states that when grown to maturity they become in their turn larva-bearing, and go into the water to bring forth. Their young are more than two (3 to 5 being the numbers observed) with a length of 33–40 mm. or of 21–23 mm. at birth. They are light grey, spotted (mottled with lighter and darker colour), have relatively short gills (8 to 9 mm. at most) and a broad tail-fin (3 mm. wide). At metamorphosis they are relatively long (44 mm.) and one of them had some yellow pigment.

Here again this summary is, as a matter of fact, describing the behaviour of two mothers, of which one produced three, and the other five young.

To my mind these experiments suggest that the reproductive habits of both species, if closely observed, will be found to be subject to considerable variation, and I think it not impossible that each species is, especially in confinement, capable of being a good deal deflected from its normal behaviour. Moreover, there seems to me no great improbability in the idea that there is an interdependence between the number of young and the stage of maturity in which they are born. But, at the same time, the case as told by Kammerer strikes me as proving too much. If each species is so sensitive to conditions that the normal procedure is gravely modified in one generation, and if that modification can reappear in a pronounced form in the next generation without a renewal of the disturbing conditions, it becomes extremely difficult to understand how the regularity which each species is believed to display in nature can be maintained. Surely both species might be expected to be in confusion. From a passage in Kammerer's earlier paper (1904, p. 55) on the subject, I infer that he also would expect considerable irregularity in the natural behaviour, but that he has not investigated the point.[20]

[19] Throughout Kammerer's papers this is used almost as a technical term. It means, I presume, that the feature was manifested more than once.
[20] It should be stated that the papers contain a quantity of detail, especially

3. *Modification of the Colour of Salamandra maculosa induced by Change in the Colour of the Soil on which the Animals were kept.*—Kammerer speaks of this as the most convincing of all his experiments on the transmission of acquired characters. So far, however, no full account of them has been published.[21] The statement is that when salamanders are kept in yellow surroundings the yellow markings gradually in the course of years increase in amount relatively to the black ground colour. Conversely by keeping the animals on black garden soil, the yellow may be greatly diminished in quantity until it largely disappears. (The account in *Natur* adds that very moist conditions also favour the increase of yellow, and that with less moist conditions the yellow diminishes.) From each kind, the (induced) yellower and the (induced) blacker, a second generation was raised, on soil of neutral colour, and each family was later divided into two parts, half being put on black and half on yellow ground.

As regards the offspring of those which had lived on *black* soil no positive result had been reached up to the date of publication, but it is stated that these young resembled their parents in having the yellow distributed in *irregular spots*.

As regards the offspring of those which had lived on yellow soil the account follows up the story of that part of the offspring which were put on yellow soil again. It is stated that these, though derived from parents with irregular spots, *developed the yellow as longitudinal bands*.

This account is given with slight differences of expression in the three places to which I have referred. On returning from Vienna in 1910 I consulted Mr. G. A. Boulenger in reference to the subject, and he very kindly showed me the fine series from many localities in the British Museum, and pointed out that in nature the colour-varieties can be grouped into two distinct types,

descriptive of the state of the larvae, which I have not attempted to represent, but the account here given contains all that seemed essential to an understanding of the more important features of the account.

[21] The first appeared in *Natur*, 1909–10, Heft 6, p. 94; and the second, which contains coloured plates of the animals, in the lecture already referrred to, *12 Flugschr. d. Deut. Ges. f. Züchtungkunde*, Berlin, 1910, p. 26. In the paper in *Mendel Festschrift*, 1911, the subject is continued, but no more is added as to this part of the experiment.

one in which the yellow of the body is irregularly distributed in spots and one in which this yellow is arranged for the most part in two longitudinal bands which may be continuous or interrupted. *The spotted form is, as he showed me, an eastern variety, and the striped form belongs to western Europe.* Mr. E. G. Boulenger[22] has since published a careful account of the distribution of the two forms. The spotted he regards as the typical form, var. *typica*, and for the striped he uses the name var. *taeniata*. The typical form occupies eastern Europe in general, including Austria and Italy, extending as far west as parts of eastern France. The var. *taeniata* is found all over France, excepting parts of the eastern border, Belgium and western Germany, Spain and Portugal. Of the very large series examined there was only one specimen (Lausanne) which could not with confidence be referred to one or other of the two varieties. Mr. E. G. Boulenger points out that both varieties inhabit very large areas, and live on soils of most different colours and compositions. Both are liable to variations in the amount and the shade of the yellow, but that any suggestion that *taeniata* belongs especially to yellow soils and *typica* to black soils is altogether inadmissible. He expresses surprise that Kammerer should not allude to these peculiarities in the geographical distribution of the two forms. He suggests further that it is more likely that some mistake occurred in Kammerer's observations than that the east European *typica* should, in the course of a generation, have been transformed into the west European *taeniata* by the influence of yellow clay soil.

In his last paper on the subject Kammerer states incidentally[23] that he has found the *striped form recessive to the spotted*. No evidence for this statement is given, and I have not found any other reference to crosses effected between the two natural types. If, however, this representation is correct, it is conceivable that the production of *taeniata* from *typica* was in fact the re-appearance of a recessive form. The plate which Kammerer gives in illustration of his modified parent figures a single animal at four stages, and though it is certainly more like the spotted than

22 E. G. Boulenger, *Proc. Zool. Soc.*, 1911, p. 323.
23 *Mendel Festschrift*, 1911, p. 84.

the striped form, it has a certain suggestion of the striped arrangement, such as I can well imagine being produced in the heterozygote.[24]

In continuation[25] of the experiments on the colour of *S. maculosa* Kammerer publishes an account of elaborate experiments in grafting ovaries of the various forms, modified and unmodified, into each other, and describes the offspring which followed. Before pursuing this part of the inquiry I am disposed to wait until the earlier steps have been made much more secure than they yet are.

More recently Kammerer has published similar statements in regard to the inheritance of characters induced in various lizards by keeping them in abnormal temperatures, high and low. The changes induced affected in some species the colours, in others the reproductive habits. Respecting these examples I feel the same scepticism that I have indicated in regard to the others, somewhat heightened by the fact that insufficient evidence is given both regarding the behaviour of these various species in captivity when not subjected to abnormal temperatures, and in the wild state.

Respecting this part of the evidence Mr. G. A. Boulenger has lately published a criticism[26] from which I extract the following passages. Referring to a previous note[27] on the question of the melanism of the various insular forms of *Lacerta muralis* he writes: " I also alluded (*l. c.*) to the theories that have been propounded to explain the melanism of various insular forms. This is a subject which has been lately taken up by Dr. Kammerer at the Biologische Versuchsanstalt in Vienna, and he claims to have produced nigrinos artificially by a very strong elevation of the temperature, accompanied by extreme dryness. Dr. Werner[28] has already opposed his own experiments to those of Kammerer, artificial melanism having been produced by him in *Lacerta oxycephala* by keeping two very light specimens from

[24] *12 Flugschrift. Deut. Ges. Züchtungskunde,* 1910, Fig. 15, *P. Reihe.*
[25] *Mendel Festschrift,* 1911, p. 83.
[26] Field, 1912, 30 March.
[27] *Ibid.,* 1904, p. 863.
[28] *Mitth. Naturw. Ver. a. d. Univ. Wien,* 1908, p. 53.

15

Ragusa for a whole summer in very damp conditions. Neither is Kammerer's theory in accordance with the distribution of the black lizards, as pointed out by Werner. Kammerer also finds that those forms which are known to produce melanic races in a state of nature, lend themselves more readily than the others to the success of his experiments. But he shows himself misinformed when he states that the variety called *Lacerta fiumana* belongs to the category of those of which black forms are not known. He overlooks the fact, first pointed out by Scherer in 1904, and which I can confirm, that the black lizard from Melisello near Lissa in the Adriatic is unquestionably derived from the lizard from Lissa, which he correctly regards as not separable from *L. fiumana*. . . ."

" Another colour modification which Dr. Kammerer states that he obtained by raising the temperature is the assumption by the female of the typical *Lacerta muralis* of the bright red colour of the lower parts which often distinguishes the male from the female, and which was not shown by the individuals of the latter sex kept by him under normal conditions. He quotes various authorities to show that the lower parts are never red in the females, but he has omitted to consult others who say the contrary. Thus Bedriaga (1878 and 1879) remarks that a so-called var. *rubriventris* of the typical wall lizard has the lower parts red in both sexes."[29]

In reading such papers as those of Semon or Kammerer the thought uppermost in my mind is that to multiply illustrations of supposed transmission of acquired characters is of little use until some one example has been thoroughly investigated. If we had certain assurance that even a single unimpeachable case could be repeated at will, the whole matter would assume a more serious aspect. If, for instance, Kammerer were able to show us *Alytes* males with horny pads on their hands, it would be something tangible; still more, if the experiment were repeated by others until no doubt remained that the offspring of *Alytes* which had bred in water for some three generations did acquire these

[29] As to the variations of *Lacerta muralis* in Western Europe and North Africa see Boulenger, G. A., *Trans. Zool. Soc.*, 1905, vol. XVII, p. 351.

pads and that they could transmit these novelties to descendants raised in normal conditions. Till evidence of this kind is published by at least two independent observers investigating similar material, I find it easier to believe that mistakes of observation or of interpretation have been made than that any genuine transmission of acquired characters has been witnessed.

Meanwhile there is no denying that the origin of adaptational features is a very grave difficulty. With the lapse of time since evolutionary conceptions have become a universal subject of study that difficulty has, so far as I see, been in nowise diminished. But I find nothing in the evidence recently put forward which justifies departure from the agnostic position which most of us have felt obliged to assume.[30]

APPENDIX TO CHAPTER IX.

Professor G. Klebs, as is well known to students of evolutionary phenomena, has for several years been engaged in investigations relating to the inheritance of acquired characters. In his many publications on the subject the issue has always been represented as more or less uncertain.

Desiring to know how the matter now stands according to Professor Klebs' present judgment I wrote to him asking him to favour me with a brief general statement. This he most kindly sent in a letter dated 8th July, 1912.

As such a statement will be read with the greatest interest by all who are watching the progress of these studies I obtained permission to publish it as follows:

8. Juli 1912

Ihre liebenswurdige Anfrage will ich sehr gern beantworten, obwohl ich sie nicht so beantworten kann wie ich erwünschte. Ihr Skepticismus in der Frage der Uebertragung erworbener Charactere auf die Nachkommen ist nur zu berechtigt. Meine Versuche mit Veronica sind *nicht* beweisend, da es mir bisher nicht gelungen ist eine einigermasse konstante Varietät mit verlaubten Inflorescenze zu erzeugen. In Bezug auf mein Semper vivum bin ich allerdings noch heute der Meinung dass

[30]As to the experiments of Klebs relating to the transmission of acquired characters, see Appendix.

die starke künstliche Veränderung der Blüte einen Einfluss auf einzelnen Nachkommen gehabt hat. Ich habe seither nichts darüber veröffentlicht: die Mehrzahl der anormalen gefüllten Blüten war leider steril. Von einem weniger veränderten Exemplar erhielt ich einige Sämlinge, aber sie haben noch nicht geblüht. Es kann sich in diesem Falle nur um eine *Nachwirkung in der ersten Generation* handeln, vergleichbar jenen Fällen in denen Samen von Bäumen aus den hohen Alpen in der Ebene gewisse Nachwirkungen zeigen. Aber es ist bisher kein sicherer Fall bekannt in den der kunstliche herbeigeführte Charakter *mehrere Generationen hindurch unter der gewöhnlichen "normalen" Bedingungen* übertragen worden ist.

Auf der andere Seite sind diese negativen Resultaten nicht entscheidend. Denn wie wenig ist in dieser Beziehung überhaupt ernstlich versucht worden! Und zweifellos geht die Sache nicht so einfach.

Ich versuche es mit anderen Pflanzen weil ich der Meinung bin dass es möglich sein müsse wenigstens solche neuen Varietäten zu erzeugen, wie sie die Gartenvarietäten entsprechen.

Aber bis jetzt leider sind die Versuche nicht gelungen, weder mir noch irgend einem anderen.

CHAPTER X

EFFECTS OF CHANGED CONDITIONS CONTINUED

THE CAUSES OF GENETIC VARIATION

IN the last chapter we examined some of the evidence offered in support of the belief that adaptation in highly organised forms is a consequence of the inheritance of adaptative changes induced by the influence of external conditions. The state of knowledge of this whole subject is, as I have said, most unsatisfactory, chiefly for the reason that in none of the cases which are alleged to show a positive result have two observers been over the same ground, or as yet confirmed each other. In the wider consideration respecting the causes of variation at large we find ourselves still in the same difficulty. The study has thus far proved sadly unfruitful. In spite of the considerable efforts lately made by many observers to induce genetic variation in highly organised plants or animals, and though successes have occasionally been announced, I do not know a single case which has been established and confirmed in such a way that we could with confidence expect to witness the alleged phenomena if we were to repeat the experiment. Abundant illustrations are available in which individuals exposed to novel conditions manifest considerable changes in characters or properties, but as yet there is no certain means of determining that germ-cells of a new type shall be formed.

Of the direct effect of conditions the lower organisms, especially bacteria, offer the best examples, the alterations of virulence which can be produced in so many distinct ways being the most striking and familiar. That attenuation of virulence can be produced by high temperatures or by exposure to chemical agents, and that this diminution in virulence may remain permanent is, from our point of view, not surprising; but the fact that in many cases the full virulence can by suitable cultivation be restored is difficult to understand. Similar variations have been observed in power of pigment production and other properties.

These phenomena naturally raise the question whether any cases of apparent loss of factors in higher forms may be comparable.

The subject of variations in the lower organisms and their dependence on conditions is a highly special one, and I have no knowledge which can justify me in offering any discussion of them, but I understand that hitherto little beyond empirical recognition of the phenomena has been attempted. A useful summary of observations made by many investigators was lately published by Hans Pringsheim,[1] who enumerates the different agencies which have been observed to produce modifications, and the various ways in which these changes are manifested. One of the most comprehensive studies of the subject from the genetic point of view is that made by F. Wolf.[2] In his extensive cultivations of *Bacillus prodigiosus*, *Staphylococcus pyogenes* and *Myxococcus* he succeeded in producing many strains with modified properties. In most of these the modifications arose in consequence of the application of high or low temperatures or of the addition of various chemical substances to the culture-media. Some of the variations, which are for the most part in the powers of pigment-formation, persisted when the strains were returned to normal conditions, and others did not. In reference especially to the variations witnessed in the Cocci the reader should consult the critical account of variation in that group published by the Winslows,[3] where much information on the subject is to be found. The authors attempted to determine the systematic relationships of the several forms, as far as possible, by the application of statistical methods. The result is interesting as showing that the problem of species in its main features is presented by these organisms in a form identical with that which we know so well in the higher animals and plants, whatever

[1] Pringsheim, H., *Die Variabilität niederer Organismen*, Berlin, 1910.

[2] F. Wolf, Modifikationen u. Mutationen von Bakterien, *Zts. F. indukt. Abstam. u. Vererbungslehre*, II, 1909, p. 90.

[3] Winslow, C. E. A. and A. R., *Systematic Relationships of the Coccaceae*. New York. 1909.

properties be selected as the diagnostic characters. There are many types perfectly distinct and others which intergrade. Some of the types change greatly with conditions while others do not. This is exactly what we encounter whenever we study the problem of species on an extended scale among the higher forms of life.

There is now practically complete agreement among bacteriologists that the observations made first by Massini on the change in color of *Bacterium coli mutabile* grown in Endo's medium, associated with the acquisition of the power to ferment lactose, are perfectly reliable and free from possibilities of mistake. The work has been extended and confirmed by many workers, especially R. Müller, who finds that this bacterium can similarly acquire and maintain the power to ferment other sugars. A careful account of the whole subject written by Müller for the information of biologists will be found in *Zts. für Abstammungsl.*, VIII, 1912. After discussing the biological significance of the facts, he concludes with a caution to the effect that bacteria are so different from all other living things that generalizations from their behavior must not be indiscriminately applied to animals and plants.

In all work with this class of material there is obviously danger of error through foreign infection of the cultures, but there can be no doubt that though some of the "mutations" recorded may be due to this cause, the majority of the instances observed under stringent conditions are genuine.

Another and equally serious difficulty besetting work with bacteria and fungi cultivated from spores is that the appearance of variation may in reality be due to the selection of a special strain previously living masked among other strains. This possibility must be remembered especially in those instances which are claimed as exemplifying the effects of acclimatisation. Manifestly this consideration can be urged with most force when the strain which gave rise to the novelty was not raised from a single individual spore. Moreover, when once the possibility of spontaneous variation is admitted, it must be difficult to be quite confident that any given variation observed is in reality due to the novel conditions applied, and as I understand the evidence, the appearance of the mutational forms does not with any regularity follow upon the application of the changed conditions.

Researches into the variation of these lower forms will, no doubt, be continued on a comprehensive scale. So long as the instances recorded are each isolated examples it is impossible to know what value they possess. If they could be coordinated in such a way as to provide some general conception of the types of variation in properties to which bacteria, or any considerable group of them, are habitually liable, the knowledge might greatly advance the elucidation of genetic problems.

Of mutational changes directly produced with regularity in micro-organisms by treatment, the experiments with trypanosomes provide some of the clearest examples. A summary of the evidence was lately published by Dobell,[4] from which the present account is taken. The most definite fact of this kind established is that certain dyes introduced into the blood of the host have the effect of destroying the small organ known as the "kineto-nucleus" in the trypanosomes. The trypanosomes thus altered continue to breed, and give rise to races destitute of kinetonuclei. This observation was originally made by Werbitzki and has been confirmed by several observers. The exact way in which this alteration is effected in the trypanosomes is not quite definitely made out, but there is good reason for supposing that the dyes have a direct and specific action upon the kineto-nucleus itself, and circumstances make it improbable that in some division a daughter-organism without that body is produced, or that any selection of a pre-existing defective variety occurs.

Ehrlich has suggested with great probability that the dyes which possess this action owe it to the fact that they have the particular chemical linkage which he calls "ortho-quinoid." In outward respects, such as motility and general appearance, the modified organisms are unchanged, but their virulence is diminished. As regards the possibility of the defective strain re-

[4] C. C. Dobell, *Jour. Genetics*, 1912, II, p. 201, where full references are given. Still more recently the same author has contributed an excellent summary of the evidence relating to bacteria (*ibid.*, II. 1913, p. 325).

acquiring the kinetonucleus, Werbitzki states that in one case passage through 50 animals and treatment with dyes left the strain unaltered; but that in another case at the sixteenth passage 7 per cent. of the trypanosomes were found to have re-acquired the organ, and in subsequent passages the percentage increased, until at the twenty-seventh passage practically all had re-acquired it.   Kudicke, however, in similar experiments did not succeed in causing re-acquisition by transplantation.

By the action of various drugs and antibodies races of trypanosomes resistant to those substances have been obtained. These breed true, at least when kept in the same species of animal in which the resistance was acquired.   As to whether change of virulence is produced by passage through certain animals or not, there is as yet no general agreement.

Other changes, especially in size and some points of structure, are said to occur when certain trypanosomes proper to mammals are passed through cold-blooded vertebrates (Wendelstadt and Fellmer), and it is stated that these changes persist, but the observations have not yet been confirmed.

Experiments lately conducted by Woltereck with *Daphnia* are interesting as having given a definite positive result, in so far, at least, as the ova were affected by conditions before leaving the bodies of the parent individuals.   The observations relate to the offspring resulting from *parthenogenetic* eggs.   Females bearing ephippia (fertilised eggs) were isolated until the ephippia were dropped, and in this way the offspring of fertilisation were excluded.   Males, of course, appeared from time to time in the cultures, but as fertilised eggs were rejected, their presence did not disturb the result.   The most remarkable observations related to *Daphnia longispina.*

This species as found in the lower lake at Lunz had the front end of the body blunt and nearly round in profile; but on being cultivated in a warm temperature and with abundant nourishment the front end of the body became produced into an elongated "helmet," as Woltereck calls it.   Experiment showed that the change was primarily due to the abundance of food, and owing to temperature in a subordinate degree.

This distinction arose as soon as the species was taken into the hothouse, but when the modified individuals were put back into the original conditions, a lower temperature and scanty food-supply, the next generation returned to their original form. After being cultivated for two years and about 40 generations in the more favourable conditions, when similarly put back into the lower temperature with scanty food the *first generation* born in these conditions was helmeted like the modified parents. Woltereck is of opinion that the ova were still unformed at the time the parents were put back, and the influence of the favourable conditions upon the unformed ova he speaks of as a "praeinduction." The effect never extended beyond the one generation, after which the strain returned to its original state.

The fact that the influence on the offspring was not manifested at first led Woltereck to expect that by more prolonged cultivation in the favourable conditions a further extension of this influence would be produced, but this expectation was never fulfilled, though the attempt was made again and again.

Similar experiments were made with *Hyalodaphnia cucullata*, which is far more sensitive to cultural influences, and in nature manifests a considerable elongation of the helmet as a seasonal modification, but the results were essentially the same as in the preceding case, no modification extending beyond the first generation born after the restoration to *normal conditions*.[5]

The only criticism of these extremely interesting results which suggests itself is that perhaps the original appearance of the modification was not in reality due to an *accumulated* effect of the conditions, but to some change in the conditions themselves which was not noticed. It is difficult to see how length of time or even the lapse of several generations could have so specific an effect on the race. It is no doubt often vaguely supposed by many that a long period of time may be necessary for the effect of climate or of other environmental conditions to be produced in an organism which does not thus respond at first. I have never been able to see any reason for this opinion nor how

[5] See Woltereck, *Verh. d. Deut. Zool. Ges.*, 1909, p. 110; and 1911, p. 142. This is a subject which can only be properly appreciated on reference to the original papers. Several complications are involved to which I have not here alluded.

it is to be translated into terms of physiological fact, and I imagine that in those cases in which the lapse of time is really required for the production of an effect, the influence of the prolongation is rather on the conditions than on the organisms. The response of the organisms thus probably indicates not that the creature is at length feeling the effects because of their accumulated action on itself, but that the conditions have at length ripened.

As this sheet is passing through the press Agar has published[6] an abstract of evidence as to another comparable case in a parthenogenetic strain in the daphnid, *Simocephalus vetulus*. When fed on certain abnormal foods the shape of the body is changed, the edges of the carapace being rolled backwards so as to expose the appendages. The offspring of animals thus modified showed similar modification in the first, and to a very slight degree, in the second generation, though the original mothers were removed to normal conditions before their eggs were laid. In the third generation there was "a very pronounced reaction in the opposite direction." Agar suggests that the change may be due to some toxin-like substances, carried on passively by the egg into the next generation, against which the protoplasm eventually produces an anti-body.

The experiments which have been in recent years regarded by evolutionary writers as the most conclusive proof that direct environmental action may produce germinal variation are those of Professor W. L. Tower, of Chicago, on *Leptinotarsa*, the potato beetles. This work has attained considerable celebrity and has been generally accepted as making a definite extension of knowledge. After frequently reading Tower's papers and after having been privileged to see some of the experiments in progress (in 1907) I am still in doubt as to the weight which should be assigned to this contribution.

The work is described in two chief publications, the first of which appeared in 1906.[7] This treatise contains a vast amount of information about numerous species and varieties of these

---

[6] *Proc. Roy. Soc.*, B, Vol. 86, 1913, p. 113.

[7] *An Investigation of Evolution in Chrysomelid Beetles of the Genus Leptinotarsa*, Carnegie Publications, 1906, No. 48.

beetles which the author has observed and bred in many parts of their distribution throughout the United States, Mexico and Central America. The part of the book which has naturally excited the greatest interest is that in which Tower states that by subjecting the beetles to change in temperature and moisture, he caused them to produce offspring quite unlike themselves, which in several cases bred true.

It is much to be regretted that the author did not happen to become acquainted with Mendelian analysis at an earlier stage in the investigation. The evidence might then have been handled in a much more orderly and comprehensive way, and a watch would have been kept for several possibilities of error.

The headquarters of the genus is evidently as Tower states, in Mexico and the adjoining countries. In this region there is a great profusion of forms, some very local, some as for instance the well-known *decemlineata*,[8] more widely spread. The distinctions are almost all found in peculiarities of colour and pattern, and the limits of species are even more indefinable than is usual in multiform animals. Tower arranges the various types into seven groups of which the one most studied is that which he calls the *lineata* group. To this group belong all the forms to which reference is here made, and, as I understand, they differ among themselves entirely in size, colour and pattern. There is no suggestion of infertility in the crosses made between the several forms of the *lineata* group; in fact they present, like many Chrysomelidae, a good example of what most of us would now call a polymorphic species, consisting of many types, some found existing in the same locality, others being geographically isolated.

A series of experiments was devoted to the attempt to fix strains corresponding to the extremes of continuous variations. For example, those with most black pigment and those with least black taken from a population continuously varying in this respect, were separately bred; but almost always the selection led to no sensible change in the position of the mean of the popu-

----

[8] This is the famous Colorado beetle or potato-bug, which has caused such serious destruction in potato crops. There seems to be no doubt that this insect, formerly unknown in the eastern States, made its way east along the mining trails when the west was opened up.

lation. The variations in these cases were evidently fluctuational. In some instances, however, real genetic differences were met with, and strains exhibiting them were, as usual, rapidly fixed.

Tower points out that several of the varieties (or species, as he prefers to call them) were obviously recessive to *decemlineata*. This is most clearly demonstrated in the case of the form called *pallida*, which is a pale depauperated-looking creature, with the orange of the thorax almost white and the eyes devoid of pigment.[9] This form behaved as an ordinary Mendelian recessive, breeding true whenever it appeared in the cultures, or when individuals found wild were studied in captivity. A black form which Tower names *melanicum* was similarly shown to be a Mendelian recessive. Wild specimens of this variety of opposite sexes were not found simultaneously in nature, and there was thus no opportunity of breeding them together, but the hereditary behaviour was seen in the F_2 generation from a *melanicum* found coupled with *decemlineata*. Experiments also occurred giving indication that a variety with the stripes anastomosing in pairs (*tortuosa*), was another recessive, and that a variety—called "*rubri-vittata*"—gave an intermediate F_1 with subsequent segregation. All these are forms of *decemlineata* Stål.

Similar observations were made regarding forms recessive to *multitaeniata* Stål. Of these two were thrown by *multitaeniata* itself, namely a form named by Stål *melanothorax*, and regarded by him as a species, and one which Tower names *rubicunda* n. sp. The facts proving the recessive behaviour of their several forms will be found in the following places in Tower's book:

pallida, pp. 273–278.
melanicum, p. 279.
tortuosa, p. 280.
rubrivittata, pp. 280–281.
melanothorax and *rubicunda*, pp. 283–285.

Following this evidence of recessive nature of the six forms

[9] This is indicated in the coloured plate, but I have not found any explicit statement to this effect in the text, and am not sure if the absence of pigment was regarded as complete.

enumerated, Tower describes experiments showing, as he believes, that some of them may be caused to appear by applying special treatment to the parents during the "growth and fertilisation" (p. 287) of the eggs. The most striking example is that in which 4 males and 4 females of *decemlineata* were kept very hot (average 35° C.) and dry, and at low atmospheric pressure (19–21 inches). The eggs laid were restored to natural conditions. These gave 506 larvae, from which emerged 14 normal, 82 *pallida* and 2 "*immaculo-thorax*," viz., without pigment on the pronotum. The account of the rest of the experiment is somewhat involved, but I understand that the *pallida*, of which two only survived, behaved as normal recessives when bred to the type: also that the parents, after having laid the eggs whose history has been given, were restored to normal conditions and laid 319 eggs which gave 61 normals.

In another case normal parents laid 409 eggs in the hot and dry conditions, and on restoration to normal conditions, the same parents laid 840 eggs. Then 409 eggs gave 64 adults as follows:

	Males	Females
decemlineata	12	8
pallida	10	13
immaculothorax	2	3
albida	9	7
	33	31

The 840 eggs laid in normal conditions gave 123 normal *decem-lineata*.

Similar experiments were made with *multitaeniata* and gave comparable results, the two recessives (*melanothorax, rubicunda*) being produced in large numbers when the parents were subjected to heat, but in this case the atmosphere was kept *saturated* with moisture, instead of dry, as in the previous instance. The same parents transferred to normal conditions gave normals only.

Lastly the form *undecimlineata* was exposed "to an extreme stimulus of high temperature, 10° C. above the average," and a dry atmosphere, with the result that from 190 eggs there emerged 11 beetles, all of the form *angustovittata* Jacoby, which subsequently bred true to that type (see p. 295).

In the results of these experiments, as described, there is one feature which I regard as quite unaccountable. Tower makes no comment upon it. Indeed, from the general tenour of the paper, I infer, not only that he does not perceive that he is recounting anything contrary to usual experience, but rather that he regards the result as conforming to expectations previously formed. The point in question is the genetic behaviour of the dominant normals produced under the abnormal conditions. These normals were the result of the breeding of parents declared to be at the same time giving off many recessive gametes. Some of these normals must be expected therefore to be heterozygous unless some selective fertilisation occurs. Nevertheless in every case they and their offspring are reported to have continually bred true. I allude especially to the tables given on pp. 288, 289, 292, and 293. Tower does not mention any misgiving about this result, and I think he regards himself as recounting phenomena in general harmony with the ideas of mutation expressed by De Vries. This they may be; but to anyone familiar with analytical breeding the course of these experiments must seem so surprising as to call for most careful, independent confirmation.

In 1910[10] Tower published an account of further experiments with *Leptinotarsa*. The work described related to two subjects. Crosses were made between three forms, *undecimlineata* Stål, *signaticollis* Stål and "*diversa*" named by Tower as a new species. The distinctions between these three depend partly on characters of the adults and partly on those of the larvae. The adults of *undecimlineata* and *diversa* have the elytra striped, but the elytra of *signaticollis* are unstriped. The larvae of *signaticollis* and of *diversa* are yellow, but those of *undecimlineata* are white.[11] Moreover, in *signaticollis* and *diversa* the black increases in the third stage of the larvae to form transverse bands which are absent in *undecimlineata*. The general course of the experiments shows that these differences may be approximately

[10] *Biol. Bull.*, XVIII, 1910, p. 285.

[11] This description does not quite agree with the representation of the larvae in Pl. 17 of the book *Evolution in the Genus Leptinotarsa* for there the larva of *undecimlineata* is shown as white in the second stage, but yellowish in the third stage; perhaps there is an error in printing.

represented as due to the action of three factors, any of which may be independently present or absent. The stripings of the elytra and of the larvae are each due to a separate factor. As regards the distinction between the yellow and the white larvae the evidence does not prove that there is decided dominance of either colour and I infer that the heterozygotes are often intermediate.

The chief contribution which this new paper claims to make relates to differences in the results which ensue from crosses effected between these three types at different average temperatures.

We are first concerned with four experiments which I number (1), (2), (3), (4):

1. *Signaticollis* ♀ × *diversa* ♂ bred at an average temperature of 80° F. by day and 75° F. by night, gave two groups in about equal numbers. The first (49) was pure *signaticollis* and bred true. The second (53) was of an intermediate type, which on being bred together gave the typical Mendelian result—1 *sig.*: 2 intermediate: 1 *div.*

2. Next, as the account originally stood in the published paper, we are told that *sig* ♀ × *div* ♂ bred together at a daytemp. average 75° F. and night average 50° F. gave an *intermediate* only, which subsequently produced a normal 1 : 2 : 1 ratio. The two crosses were repeated eleven times with identical results.

In a further experiment (3) *signaticollis* ♀ × *diversa* ♂ were bred under the same conditions as those used in expt. (1). They again gave *sig.* and intermediates as before in fairly equal numbers. The *sig.* as before bred true, and the intermediate gave 1 : 2 : 1, all exactly as in expt. (1).

In expt. (4) *the same parents* used in (3) were again mated under conditions of expt. (2) at the lower temperature, and this time gave *signaticollis* exclusively, which bred true for four generations. This experiment was repeated seven times with uniform results.

Diagrams are given representing all these histories in graphic fashion.

From these observations, Tower concludes that the determination of dominance, and the ensuing type of behaviour, is clearly a function of the conditions incident upon the combining germ plasms.

It will be observed that expts. (1) and (3) gave identical results but (2) and (4), though much the same conditions were applied, are at variance, for (2) gave all intermediates, while (4) gave all *signaticollis*. In *Amer. Nat.*, XLIV, 1910, p. 747, Professor T. D. A. Cockerell commented on this paper of Tower's and pointed out that there must be an error somewhere, for when he discusses these experiments Tower speaks of (2) and (4) as confirming each other. To this Tower replied[12] that there had been a mistake. He states that in preparing the paper "certain minor experiments were taken from a larger series and combined to illustrate a general point in the behaviour of alternative characters in inheritance," and that expt. (2) was introduced inadvertently in place of another which he desires to substitute. In this, which I number (5), *signaticollis* ♀ × *diversa* ♂ from exactly the same stocks as those used in (1), were mated at the lower temperatures specified for (2), day average 75° F., night average 50° F. These gave all of the *signaticollis* type with a narrow range of variability, which bred true, in some cases to F_6. Tower says he has repeated this experiment six times with identical results.

Nevertheless he proceeds to say that the description of expt. (2), which was repeated eleven times with identical results, was correct "as far as given." That experiment was "from a second series of cultures parallel to the one given, but in which there are other factors involved, which in H. 410 [my (2)] are productive of a typical Mendelian behaviour." He adds he does "not care at this time to make any statement of what these factors are, nor of their relations to the behaviours given in the H. 409, H. 411, H. 409/11 series [my (1), (5) and (3)–(4)] which are the simplest and most easily presented series obtained in the crossing of *signaticollis* and *diversa*."

Professor Cockerell's intervention has thus elicited the fact

[12] *Biol. Bull.*, XX, 1910, p. 67.

16

that we have as yet only a small selected part of the evidence before us, even as concerning the effect of temperature on the cross between *signaticollis* ♀ × *diversa* ♂. We learn that at the lower temperatures the result was eleven times the expected one, and six times an unexpected one; further, that we owe it to the author's inadvertence that we have come to hear of the expected result at all, and that though he knows the factors which determine the discrepancy, he declines for the present to name them. In these circumstances we can scarcely venture as yet to estimate the significance of these records.

The paper goes on to recount somewhat comparable, but more complex instances in which the descent of the colour of adults and of larvae was affected by temperature in crosses between *undecimlineata* and *signaticollis*. As they stand the results are very striking and unexpected, but I think, in view of what has been admitted respecting the former part of the paper, full discussion may be postponed till confirmation is forthcoming.

One feature, however, calls for remark. This second paper is written apparently without any reference to the discoveries related by Tower in his previous book, to which no allusion is made. This is most noticeable in the case of an experiment in which (p. 296, H. 700A) *undecimlineata* ♀ (the dominant) was mated to *signaticollis* ♂ with the result that all the offspring were *undecimlineata* and bred true to that type (Parthenogenesis was tested for, but never found to occur). This experiment was made at a temperature averaging 95° F. ± 3.5° by day and 89° F. ± 4.8° by night, and in a humidity given as 84 per cent. by day and 100 per cent. by night; but in the previous book (p. 294) we are told that pure *undecimlineata* bred together "under an extreme stimulus of high temperature, 10° C. above the average" and a relative humidity of 40 per cent. gave 11 beetles only, all *angustovittata*. But reference to the Plate 16, Fig. 2, shows that *angustovittata* must be exceedingly like *signaticollis*, having, like it, the elytral stripes obsolete, and if there is any marked difference at all, it can only be in the larvae. It seems strange that if *undecimlineata* really gives off ova of this recessive type at high temperatures, the fact should not be alluded to in connection with expt. H. 700A, where, as the father was *signaticollis*, having the same recessive character, their appearance might have been expected not to pass unobserved. The temperature in the older experiment is, of course, not given with the great accuracy used in the second, and it may have been higher still. The humidity also was widely different. Still, in discussing the phenomena we should expect some reference to the very remarkable and closely cognate discovery which Tower himself had previously reported in regard to the same species.[13]

The hesitation which I had come to feel respecting these two publications of Tower's has been, I confess, increased by the appearance of a destructive criticism by Gortner[14] who has examined the parts of Chapter III of Tower's book, in which he discusses at some length the chemistry of the pigments in *Leptinotarsa* and other animals. As Gortner has shown, this discussion, though offered with every show of confidence, exhibits such elementary ignorance, both of the special subject and of chemistry in general, that it cannot be taken into serious consideration.

Some observations made by Dr. W. T. Macdougal[15] have also been interpreted as showing the actual causation of genetic variation by chemical treatment. Of these perhaps the least open to objection were the experiments with *Raimannia odorata*, a Patagonian plant closely allied to *Oenothera*. The ovaries were injected with various substances and from some of the seeds which subsequently formed in them a remarkable new variety was raised. This varying or mutational form was strikingly different from the parental type, with which it was not connected by any intergradational forms, and it bred true. It made

[13] As to the interrelations of these three forms, Tower states (1906, p. 18) that *angustovittata*, which he reared from *undecimlineata*, is intermediate between it and *signaticollis*. Compare Stål, "*Monogr. des Chrysomélides*," 1862, p. 163; and Jacoby, *Biol. Centr. Amer. Celeopt.*, vi, Pt. 1, p. 234, Pl. xiii, fig. 20; Tab. 41, fig. 15; *ibid.*, Suppl., p. 253. All these forms are evidently very closely related, and the delimitation of species is quite arbitrary. Jacoby indeed suggests that *undecimlineata* may be a variety of *decemlineata*.

[14] Gortner, *Amer. Nat.*, Dec., 1911, XLV, p. 743.

[15] *Mutations, Variations, and Relationships of the Oenotheras*, Carnegie Institution Publication No. 81, 1907, pp. 61–64.

no rosette, growing to a much smaller size than the parent, and was totally glabrous instead of being very hairy as the parental type is. I was shown specimens of these plants by the kindness of Dr. Britton in the Bronx Park Botanic Garden in 1907 and can testify to their very remarkable peculiarities. They had a somewhat weakly look, and might at first sight be thought to be a pathological product, but they had bred true for several generations. From the evidence, however, I am by no means satisfied that their original appearance was a consequence of the treatment applied. This treatment was of a most miscellaneous description. Two of the mutants came from an ovary which had been treated with a ten per cent. sugar solution. Ten came from one into which a 0.1 per cent. solution of calcium nitrate had been injected. One was from a capsule which "had been exposed to the action of a radium pencil." Macdougal speaks of these results as decisive, but clearly before such evidence can be admitted even for consideration it must be shown by control experiments that the individual plants which threw the mutant were themselves breeding true in ordinary circumstances. Nothing is more likely than that the mutant was an ordinary recessive. I may add that Mr. R. H. Compton made a number of experiments with *Raimannia odorata*, raised from seeds kindly given me by Dr. Britton, injecting the ovaries with a variety of substances, including those named by Macdougal; but though a numerous progeny was raised from the ovaries treated, all were normal. Macdougal relates also that some mutational forms came from ovaries of *Oenothera Lamarckiana* exposed to radium pencils, and also from *Oenothera biennis* injected with zinc sulphate a peculiar mutant was raised, but taking into account the frequency of these occurrences in those species, he very properly regarded this evidence as of doubtful application. In a later paper,[16] however, he has returned to the subject and affirms his conviction that the appearance of a mutant among seedlings raised from an ovary of *Oenothera biennis* treated with zinc sulphate was really a consequence of the injection, saying that

[16] Macdougal, D. T., "Alterations in Heredity induced by Ovarial Treatments," *Bot. Gaz.*, vol. 51, 1911, p. 241.

the variation previously observed in the species was afterwards shown to be due to fungoid disease. The circumstances to which he mainly points in support of his view is that the mutation bred true, but this is only evidence of its genetic distinctness, which may, of course, be admitted by those who remain unconvinced as to the original cause of its appearance. He adds that he is making similar experiments with some twenty genera; but what is more urgently needed is repeated confirmation of the original observation. When it has been shown that this mutation can be produced with any regularity from a plant which does not otherwise produce it on normal self-fertilisation, the enquiry may be profitably extended to other plants.

A curious and novel experiment, which however, led ultimately to a negative result, was made by F. Payne. Many discussions have been held respecting the blindness of cave animals. The phenomenon is one of the well-known difficulties, and most of us would admit that the theory of evolution by the natural selection of small differences does not offer a really satisfying account of it. Those who believe in the causation of such modifications by environmental influences and in their hereditary transmission make, of course, the simple suggestion that the darkness is the cause of the loss of sight, and that disuse has led to the reduction of the visual organs. Payne bred *Drosophila ampelophila*, the pomace-fly (which is easy to keep in confinement, fed on fermenting bananas), for sixty-nine generations in darkness. At the end of that period there was no perceptible change in the structure of the eyes, or in any other respect. The number of generations may possibly be regarded as insufficient to prove anything, but comparing them, as he does, with the generations of mankind, we see that they correspond with a period of about two thousand years, an interval far longer than those which many writers in particular cases have deemed sufficient.

In his first paper Payne states that, though no structural difference could be perceived, the flies which had been bred in the dark reacted less readily to light than those which had been reared under normal conditions, and he inclined to think that the treatment had thus produced a definite effect. After more careful tests, however, he withdrew this opinion. It proved that both individual flies and individual groups of flies, both of those bred in the light and of those bred in the dark, differed greatly in their reactions, which were measured by counting the time that it took for a fly to travel to the light end of a covered tube, various sources of error being eliminated. He found further that these differences of behaviour were not inherited in any simple way, but he is disposed to attribute them to accidental differences in the nature of the food, an account which seems probable enough.[17]

In several recent publications Blaringhem[18] has described the origin of many abnormal forms of plants, especially of maize, which he attributes to various mutilations practised upon the parents. Respecting these the same difficulty which has been expressed in other cases reappears, that before drawing any conclusion as to the value of such evidence we require to know that the plants treated belong to a really pure line, which if left to nature in the ordinary circumstances of its life in that locality would have had normal offspring. Abnormalities abound in the experience of everyone who examines pans of seedlings of almost any species of plant, and in maize they are well known to be exceptionally common. Some of those which we meet with when we attempt to ripen maize in this country are very similar to those which Blaringhem describes, consisting in irregularities in the distribution of the sexes, in the shapes of the panicles, etc. Many of these are doubtless imperfections of development, due to the dullness of our climate, but others are presumably genetic and would recur in the offspring however treated. If some one working in a climate where maize could be raised in perfection would repeat these experiments, and show that a strain which was thoroughly reliable and normal in its genetic behaviour did, after mutilation, throw the miscellaneous types observed by Blaringhem, that would be evidence at least that the development of the seed could be so influenced by injury to the parental tissues that its properties were changed.

[17] Payne, Fernandus, *Biol. Bull.*, XVIII, 1910, p. 188, and *ibid.*, XXI, 1911, p. 297.

[18] See especially, *Mutation et Traumatismes*, Paris, Felix Alcan, 1908.

Such evidence could be used for what it is worth; but pending an inquiry of this kind I am disposed to regard these observations of variation following on parental injury as suggestive rather than convincing.

Some evidence of a remarkably interesting kind has been collected by J. H. Powers[19] respecting the structure and habits of *Amblystoma tigrinum*, which led him to the conclusion that striking differences in the form, anatomy, and developmental processes could be effected directly by change in the conditions of life. It is well known that a profusion of forms, distinct in various degrees, is grouped round *Amblystoma tigrinum*. Some of these are believed to be geographically isolated, others occur together in the same waters, and, as usual, authorities have differed greatly as to the number of names to be given. These forms were studied in detail by Cope who described them in the *Batrachia of North America*. The view which he inclined to take was that the individual variations of *Amblystoma tigrinum* resulted from variations in the time and completeness of the metamorphosis, and these were regarded as due to external causes, such as differences in season, temperature, and geographical conditions. Powers, however, states that collecting within a radius of six or eight miles he found almost if not quite the whole "gamut of recorded variation in this species." Some, however, as he states, occurred rarely except under experimental conditions, but considerable differences in temperature were not found necessary in producing them. Every year, he says, he has been able to add to the number of peculiar types found in the same small area in nature, until the amount of natural variation at least equals that seen by Cope in the collections of the National Museum and those of the Philadelphia Academy.

Powers states that his observations by no means confirm Cope's view that these differences are in the main referable to variation in the completeness of metamorphosis, and on the contrary, he regards metamorphosis as on the whole a levelling process, tending to obliterate diversity. The enormous dif-

[19] J. H. Powers, "Morphological Variation and its Causes in *Amblystoma tigrinum*." *Studies from the Zoological Laboratory.* The University of Nebraska, No. 71, 1907.

ferences in size and proportions which he describes can only be
appreciated by reference to his figures. They affect almost all
features of bodily organisation. These striking differences he
looks upon as brought about by differences in nutrition, "diversities in habitual locomotion," and diversity in the age at which
metamorphosis occurs, and to sexual difference. Apart from
sexual difference he regards the chief distinctions, in brief, as
"acquired variations of the larva."

As an example he gives the great elongation of some of the
forms as "due first to slow growth, second to the free-swimming
habit, third to the prolongation of larval life, and finally to the
assumption of sexual maturity as males," either in the branchiate
or non-branchiate condition. He describes the rapid growth of
some and the slow growth of others. A larva of intermediate
type may grow about a centimeter a month, but a rapidly growing
specimen may grow more than four times as much. The slower
rate of growth may, he says, be induced by winter feeding, and
other treatment.[20]

When, however, he goes on to describe the influences which
he regards as exerted by the habit of freely swimming, I am led
to wonder whether after all in most of these illustrations, the
primary distinctions are not in reality genetic. "Specimens
raised in the same aquarium or in similar aquaria, side by side
with all conditions as uniform as it is possible to make them,
seldom fail to furnish striking examples of broad-headed, short-
bodied, and short-tailed types which are habitually found at the
bottom, while others, slender and elongated, are free swimmers,
and maintain themselves in almost as continual suspension and
motion as does a gold fish." Later, again, he writes, "Yet despite
the uniformity of these favourable conditions, the larvae soon
began to split up into two noticeably distinct groups, the one of

[20] In connexion with this case I would refer the reader to some remarkable
observations of Dr. T. A. Chapman on various types of larvae which he reared
from the moth *Arctia caja* (*Ent. Rec.*, IV, 1893, p. 265, and following parts). From
a single mother he raised a great diversity of forms, some which fed up rapidly
and passed through their development without assuming certain stages, and others
which were, as he called them, "laggards," moulting more times than their brethren
and developing at a much slower rate. It is greatly to be hoped that such a case
may be critically investigated by analytical breeding.

unusually compact proportions, the other of uniform intermediate build, such is most commonly met with." It is to my mind scarcely possible to resist the inference that, though there may be definite responses to certain conditions, yet the chief distinctions are genetic, and that it is these distinctions which confer the power to respond. The parts respectively played by cause and effect are always difficult to assign; but when it is stated that "a weak-limbed, long-bodied and long-tailed animal becomes well nigh perforce an undulatory swimmer, while the strong-limbed, short-tailed, heavy-bodied specimen, when these characteristics are rapidly forced upon it, is, under certain circumstances, just as forcibly induced to become a crawler," we feel how erroneous any estimates of causation are likely to be.

One of the most remarkable and interesting sections of Powers' paper is that in which he describes the differences in bodily structure and habits which he attributes to cannibalism, and the whole account of the phenomena should be read in the original. It appears that there are two extremely distinct types of larvae, those with narrow heads and slender bodies which live for the most part on small Crustacea such as *Daphnias*, and those with huge mouths and very wide heads, which disregard such small animals altogether and live on amphibian larvae, whether of their own or other species. As the illustrations show, the differences between these two types are very great, and the differences in instinct and behaviour are no less. The cannibals take no heed of the pelagic crustacea, lying sluggishly at the bottom, rousing themselves immediately to a violent attack on the larger living things which approach them. Nothing but the most incontrovertible evidence based on abundant control experiments should convince us that such differences are not primarily genetic, and in the present state of knowledge I incline to think that the families really consist of individuals which are ready to assume the cannibal habit if opportunity offers, and others which are congenitally incapable of it. It may readily be that if all chance of cannibal diet be excluded, the full development of the wide head and mouth, or the other peculiarities, would never become pronounced, but I doubt whether such change could be induced in any individual taken at random.

CHAPTER XI.

STERILITY OF HYBRIDS. CONCLUDING REMARKS.

WHEN we consider the bearing of recent discoveries on those comprehensive schemes of evolution with which we were formerly satisfied, we find that certain details of the process are more easy to imagine. We readily now understand how varieties once formed, can persist, but at the same time difficulties hitherto faced with complacency become formidable in the light of the new knowledge. So generally is this admitted by those familiar with modern genetic research that most are rightly inclined to postpone the discussion. The premisses, indeed, on which such a discussion must be based are almost wholly wanting.

The difficulties to which I chiefly refer are not those created by the phenomena of adaptation, though they are serious enough. In treating of that subject I have felt obliged to express scepticism as to the validity of nearly all the new evidence for the transmission of acquired characters. At the present time the utmost we are bound to accept is the proof that (1) in some parthenogenetic forms variations, or perhaps we may say malformations, produced in response to special conditions, recur in one or perhaps two generations asexually produced after removal to other conditions. (2) That violent maltreatment may in rare instances so affect the germ-cells contained in the parents as to cause the individuals resulting from the fertilisation of those cells to exhibit an arrest of development similar to that which their parents underwent.

I do not doubt that evidence of this type will be greatly extended. As a contribution to genetic physiology these facts are very important and interesting, but I cannot think that any one, on reflexion, will feel encouraged by such indications to revive old beliefs in the direct origin of adaptations.

In these respects we are simply left where we were. The force of objections based upon the existence of adaptive mechanisms is no greater than it has always been. On the contrary the fact that variations can now so generally be recognized as definite is some alleviation of the difficulty. We can moreover disabuse ourselves of the notion that for all characters which are definite or fixed, some utilitarian rationale may be presumed. Upon that point the study of variation has provided a perfectly clear answer.

In frankly recognizing that the fixity of characters in general need not connote usefulness to their possessors we deliver ourselves of a distracting pre-occupation and prepare our minds for an investigation of the properties of living organisms in the same spirit as that in which the chemist and the physicist examine the properties of unorganized materials. The creature persists not merely by virtue of its characteristics but in spite of them, and the fact of its persistence proves no more than that on the whole the balance of its properties leaves something in its favour.

It may be noted by the way that the fact that the structures of living things are on the whole adaptative was not always obvious. Though to naturalists of this generation it is a truism, we have only to turn to Buffon to find that in his philosophy of nature it played no essential part. The passage in which Buffon describes what he regards as the forlorn and degraded condition of the Woodpecker is well known. We have come to think of the Woodpecker as a capital example of adaptation to the mode of life; but Buffon after enumerating the hard features of the bird's existence, forced to earn its living by piercing the bark of trees in an attitude of perpetual constraint, remarks[1] " Tel est l'instinct étroit et grossier d'un oiseau borné a une vie triste et chétive. Il a reçu de la Nature des organes et des instrumens appropriés a cette destinée *ou plutôt il tient cette destinée même des organes avec lesquels il est né* " (my italics). His reflexions on the Stilt (*Himantopus*) read even more strangely to us, accustomed as we are to see in the prodigious length and thinness of the shanks and in the other features of its organisation palpable adaptations to a wading life. For Buffon, however, this

[1] Buffon, *Hist. Nat.*, Oiseaux, 1780, VII, p. 3.

curious bird seemed a poor, neglected production, extravagant in its disproportions, one of the misfits of creation, left as a shadow in the picture composed of nature's more successful efforts.[2] This theme he develops at some length, being evidently well pleased with the idea.

Our way of regarding these things is doubtless sounder and more fruitful than Buffon's, but it is well to remember that what seems so obvious to us looked quite differently to other excellent observers; and stupid as it may have been to have overlooked plain examples of adaptation, it is a far worse mistake to see adaptation everywhere. I do not seek to minimise the real and permanent difficulty which the existence of adaptations creates, but by the suggestion that all normal specific differences are adaptational that difficulty was quite gratuitously increased.

In these respects it may be claimed that progress has been made, even if that progress seem outwardly of small account.

But all constructive theories of evolution have been built on the understanding that what we know of the relation of varieties to species justifies the assumption that the one phenomenon is a phase of the other, and that each species arises or has arisen from another species either by one or several genetic steps. In the varieties we have accustomed ourselves to think that we see those steps. We still know little enough of the mode of occurrence of variation, but we do begin to know something, and if we ask ourselves whether our knowledge, such as it is, conforms at all readily with our former expectations, we cannot with any confidence assert that it does. Among the plants and animals genetically investigated are many illustrations of very striking and distinct varieties. Many of these might readily enough be accepted as species by even the most exacting systematists, and not a few have been so treated in classification; but when we have examined their relationship to each other we feel not merely that they are not species in any strict sense but that the distinctions they present cannot be regarded as stages in the direction of specific difference. Complete fertility of the results of inter-crossing is and I think must rightly be regarded as incon-

[2] Ibid., VIII, p. 115.

sistent with actual specific difference; and of variations leading to that consequence no clear indication has yet been found. As an example of possible exceptions mention should perhaps be made of the case of a giant form of *Primula sinensis* investigated by Keeble.[3] It arose from a " Star " Primula of normal size, and though fertile with its own pollen all attempts to fertilise it with the pollen of other forms failed. Miss Pellew, who did these fertilisations, tells me that very extensive trials were made, and repeated in several seasons. Ultimately two plants were raised from it fertilised with a plant of the strain from which it sprang, and these proved sterile. In the light of modern expe⁀ rience the significance of such isolated instances is doubtful.

All the strains known as " Giants " are, as Messrs. Sutton have always found, more or less sterile, and their sterility is presumably due to some negative defect.

In regard to the fertility of Primula species there are several paradoxes. For example the long-styled varieties, apart from giants, are fertile with their own pollen, and for many years short-styled plants have not been used in most strains. Auriculas and Polyanthuses, on the contrary, are generally if not always bred from short-styled plants, as the florists have decided that the long-styled are inadmissible. Mr. R. P. Gregory tells me that, though most strains of *P. sinensis* give seed enough when only long-styled plants are used, he finds nevertheless that when a " legitimate " union is made the amount of seed usually increases much as Darwin observed. Darwin's statement that plants of " illegitimate " origin are less fertile than the " legitimately " raised plants is also in general confirmed by his experience. To this rule there were some marked exceptions in strains derived from *long*-styled plants, which though illegitimate showed a high degree of fertility, but illegitimate unions between *short*-styled plants always produced comparatively sterile offspring. I have no records of the behavior of Auriculas and Polyanthuses. It would be interesting to know whether among them pure strains of short-styled plants (dominants) have appeared, and, if so, how their fertility is affected. Without

[3] Keeble, *Jour. Gen.*, 1912, II, p. 173.

much more critical data I suppose no one would nowadays be inclined to follow Darwin in instituting a comparison between the sterility of hybrids and that of illegitimately raised plants of heterostyle species.[4] It is even difficult to imagine any essential resemblance between these two phenomena, nor has evidence ever been produced to show that illegitimately raised plants have bad pollen grains, which is the usual symptom of sterility in hybrid plants and the consequence, as we believe, of failure of some essential division in the process of maturation.

The difficulty that we have no knowledge of the contemporary origin of forms, from a common stock, which when crossed together give a sterile product, is one of the objections constantly and prominently adduced from the time of the first promulgation of evolutionary ideas. In the light of recent work the objection has gathered strength. Why, if we are able to produce instances of variation colourably simulating specific difference in almost all other respects, do we never find an original appearance of this most widely spread of all specific characteristics? No doubt all breeders know that sterile animals and plants occasionally appear in their cultures, but it is more in accordance with probability that the sterility in these sporadic instances should be regarded as due to defect than that it should be thought comparable with that of the sterile hybrids. For their sterility must, by all analogy with results elsewhere seen, be attributed not to the absence of something, but to the presence and operation of complementary factors leading to the production of inhibition of division; and consistently with that interpretation, we find that when from a partially sterile hybrid comparatively fertile offspring can be raised, their comparative fertility continues in the posterity generally if not always without diminution. The distinction between these several kinds of sterility was of course not understood in Darwin's time. The comparison, for example, which he instituted[5] between the sterility of " contabescent " anthers and that of hybrids no longer holds, for at least in those cases in which the nature of contabescent anthers have been genetically investigated (Sweet Pea, *Tropaeolum*) they proved

[4] *Animals and Plants*, ed. 1, 1868, II, pp. 180–5.
[5] *Animals and Plants*, ed. 1, 1868, II, p. 165.

to be a simple recessive character. Nor can we now easily suppose that the attempt there made by Darwin to suggest resemblance between the sterility produced by unnatural conditions and that of hybrids has any physiological justification.

In regarding the power to produce a sterile or partially sterile hybrid as a distinction in kind, of a nature other than those which we perceive among our varieties, I am aware that I am laying stress on an impression which may hereafter prove false. The distinction nevertheless is so striking and so continually before the eyes of a practical breeder that he can scarcely avoid the inference that when he meets a considerable degree of sterility in a cross-bred he is dealing with something belonging to a distinct category, and not merely a varietal feature of an exceptional kind.

Besides the sterility of hybrids appeal has often been made to the phenomenon of incompatibility, in its several stages of completeness, as distinguishing species. No one doubts that incompatibility may arise from a variety of causes of most diverse degrees of importance, but though sometimes referred to as an extreme case of interspecific sterility, it is really a very different matter. In regard to one phase of this incompatibility, that associated with self-sterility, some progress has been made, and we are not wholly without experimental evidence of its being within the range of contemporary variation.

Given the outline of Mendelian teaching as to gametic differentiation and the classification of individuals in a mixed population, it seemed highly probable that what we call self-sterility must mean that the species really consisted of *classes*, some of which are capable of interbreeding with others while others are not. According to the received account every individual, though incapable of fertilising itself, was supposed to be able both to fertilise and to be fertilised by any other individual. This notion has always seemed to me a self-evident absurdity, for it would imply that there can be as many categories as individuals. Such experiments, however, as I made did certainly give results consistent with that belief. I first tried Cinerarias, which are usually self-sterile, but I found no incompatible pairs of plants. Whether I was deceived by the consequences of apogamy, or whether the pollen of certain plants may belong to more than one class I do not know. The results were confused in various ways. Usually the self-fertilised plants set little or nothing, and cross-fertilised they set fully with such uniformity that the few failures could plausibly be attributed to mistakes in manipulation or to other extraneous causes. Later de Vries announced[6] (without giving particulars) that he had proved the existence of such classes in *Linaria vulgaris;* but on making experiments with that species I again got no positive results, and I came to the conclusion that in spite of inherent improbability the conventional belief must be substantially true. At last, however, the work of Correns, lately published,[7] does definitely show that in one species, *Cardamine pratensis,* classes of individuals exist such that individuals of the same class are incapable of fertilising themselves or each other, but fertilisation made between the classes is usually completely effective. Many complications were encountered and some contradictory evidence is recorded, but the general bearing of the results was positive and indubitable.

We know far too little of this phenomenon as yet to be able to understand its significance, but I suppose we may anticipate with some confidence that it will be found to be a manifestation of dissimilarity between the male and female gametes of the same individual, comparable with that first seen in the Stocks (*Matthiola*) which throw doubles—a state of things in all likelihood to be found widely spread among hermaphrodite organisms. Whether the incompatibility between species is to be associated with that of the self-steriles also cannot be positively asserted, though it seems not unreasonable to expect that such an association will be discovered.

The case of the apple and the pear is an impressive illustration of this possibility. The two species are of course exceedingly alike in all outward respects, but nevertheless the pollen of each is entirely without effect on the other. Presumably we should

---

[6] *Species and Varieties,* 1905, p. 471.
[7] Correns, *Festschr. med.-nat. Ges. zur 84 Versamml. Deutsch. Naturf. u. Aertze. Münster i. W.,* 1912.

interpret this fact as meaning not so much that the apple and the pear are in reality very wide apart, but rather that either, each is lacking in one of two complementary elements, or that each possesses a factor with an inhibitory effect. Their incompatibility may well be of the same nature as that of the classes in *Cardamine pratensis*.

Returning now to the problem of inter-specific sterility; we note, as I have said, the absence of contemporary evidence that variation can confer on a variety the power to form a sterile hybrid with the parent species. The considerations based on this want of evidence have for a long while been familiar to all who have discussed evolutionary theories, and it is worth observing the exact reason why the difficulty strikes us now with a new and special force. In pre-Mendelian times all that was known was that some forms could freely interbreed without diminution of fertility in the product, while others could not. But now we find that, by virtue of segregation, from one and the same pair of parents, or even, in the case of hermaphrodites, from one and the same individual, offspring commonly arises howing among themselves exactly such differences as distinguish species—and very good species too. This we see happening again and again. But to forms capable of arising as brethren in one family the title species has never been meant to apply, and if we are going to use the term in application to fraternal groups we must definitely recognise that by " specific " difference is to be understood simply *difference*, without any immediate or even ulterior physiological limitation whatever. Naturally, therefore, we begin to think of the appearance of sterility in crosses as something apart, and as a manifestation which distinguishes certain kinds of unions in a very special way.

I am perfectly aware that there are gradations in the sterility of hybrids as in every other characteristic upon which it has been proposed to base specific definitions; but, as also so often happens in the matter of defining intergrading categories, the difficulty in practice is not often such as to lead to actual ambiguity. I am speaking of course of those examples which are amenable to genetic experiment. As to the rest there is complete and permanent uncertainty. But the experience of the practical breeder does, I think, on the whole, support the contention to which systematists have so steadily clung under all the assaults of evolutionary philosophers, that, though we cannot strictly define species, they yet have properties which varieties have not, and that the distinction is not merely a matter of degree.

The first step is to discover the nature of the factors which by their complementary action inhibit the critical divisions and so cause the sterility of the hybrid. Thus expressed, we see the problem of inter-specific sterility in its right place; and the question why we do not now find contemporary instances of varieties lately arisen in domestication, which when crossed back with their parents, or with their coderivatives, can produce sterile products, is perceived to be only a special case of a problem which in its more general form is that of the origin of new and additional factors.

For the requisite evidence no comprehensive search has been made, but perhaps it will yet be found. All that we can say at the present time is that the incidence both of hybrid sterility, and of incompatibility also, is most capricious; and provided that two forms have such features in common that a cross between them seems not altogether out of the question, no one can predict without experiment whether such a cross is feasible, and if feasible whether the product will be fertile, or sterile more or less completely. For instance, though probably all the British and some Foreign Finches (Fringillidae) have been crossed together, and some of these crosses, as for instance, the various Canary-mules have been made in thousands, I believe no quite clear example of a fertile hybrid can be produced. Many species of Anatidae cross readily and produce fertile hybrids: others give results uniformly sterile. Though most of the Equidae can be crossed and some of the hybrids are among the commonest of domesticated animals there is no certain record of a fertile mule. Among the Canidae the dogs, wolves and jackals all give fertile hybrids, but there is no clearly authenticated instance of a cross between any of these forms and the European fox. In spite of their close anatomical resemblance it is doubtful if the rabbit and the hare have ever interbred. Many of the wild species of *Bos* have been crossed and recrossed both with each other and with many domesticated races, but I understand that no cross with the Indian buffalo (*Bos bubalus*) has yet been successful even in producing a live calf.[8] In the genus *Primula* many hybrids are known and several of them occur in nature, but hitherto no certain hybrid between *P. sinensis* and any other species has been made, in spite of repeated attempts.

In *Nicotiana* many—doubtless all—the various forms of *N. tabacum* can be crossed together without diminution of fertility, though some are very distinct in appearance, but crosses between *tabacum* and *sylvestris* are highly sterile (in my experience totally sterile[9]), though the distinctions between them are not to outward observation nearly so great as those which can be found between the various races of *Primula sinensis*.

Recently some remarkable experiments bearing closely on these questions have been published by F. Rosen.[10] They concern the forms of *Erophila* (*Draba*) *verna*, celebrated in the history of evolutionary theory as the plants especially chosen by Alexis Jordan for the exposition of his views on these subjects.

The " species " contains a profusion of forms dissimilar in many structural characters, such as the size and shape of leaves, flowers, fruits, etc. Of these forms many grow in association. Jordan found, on experiment, that each, to the number of some two hundred, bred true, and that therefore, the conventional assumption that polymorphism of this kind must mean great contemporary variability had no foundation in fact. So far

---

[8] This is a case of a somewhat different order and I mention it partly for that reason as an illustration of the complexity which such negative instances may present. The difficulty is that though the buffalo and the zebu can breed together, the foetus is too large to be born alive. (See Ackermann *Ber. d. Ver. f. Naturk.*, Kassel, 1898, p. 69. Prof. S. Nathusius, of Halle, who has great experience in crossing Bovidae, tells me that he has always failed to cross the buffalo with other species.)

[9] In a paper to be published in the Report of the Genetic Conference, Paris, 1911, Bellair states that he obtained some partially fertile hybrids in the cross *N. sylvestris* × *tabacum*. As to the various degrees of sterility in hybrids between *Nicotiana* species see Lock, R. H., *Ann. Roy. Bot. Gardens*. Peradeniya, IV, 1909, p. 195.

[10] *Beitrage zur Biol. der Pflanzen.*, X, 1911, p. 379.

indeed is the evidence from favouring the belief that such forms are in any way transitional or indeterminate, that, as is well known, Jordan used it with every plausibility to support the doctrine of the fixity of species. To certain aspects of Jordan's work we will return later in this chapter, but the matter is in the present connection of especial interest for the reason that Rosen has lately found by experiment that some of these presumably very closely allied forms, crossed together, gave hybrids more or less sterile. In the case of the offspring of one pair of forms only (*E. cochleata* and *stricta*) was the fertility undiminished, and the various degrees of sterility found in the other crosses ranged up to the extreme infertility of the hybrids between *E. stricta* × *elata*. From this cross ten plants were bred. Of these the four strongest were chosen to breed from, but two of the four proved totally sterile; one had only bad seeds; and from the fourth a single seedling was raised which in its turn proved to be sterile. From the less sterile hybrids F_2 families were raised, with the usual experience that in this and subsequent generations the sterility diminished among extracted forms, new and true-breeding types with complete fertility being thus derived from the original cross.[11]

The production of sterility as a consequence of crossing plants so nearly approaching each other as these *Erophila* " species ", do is not a little interesting, and the fact well exemplifies the futility of the various attempts to frame general expressions as to specific properties or behaviour. Commenting on his results Rosen argues that the polymorphic group commonly called by systematists *Erophila* (*Draba*) *verna* may now be regarded as having arisen by crossing, as did his own types mentioned above. The question, however, *what* species were the original progenitors of the group cannot be answered. Rosen considers that no form which he knows satisfies the requirements,

[11] One very peculiar feature was observed, namely, that all the new forms in F_2 which were bred from came true. As I understand, this statement applied to five such new types, and they were represented by 76 individuals in F_3, but further details on this point are desirable. Another curious fact was observed, namely that one of the F_1 forms (*cochleata* × *radiata*) when fertilised by *cochleata* gave a highly polymorphic family, but fertilised by *radiata* the resulting offspring were almost uniform.

and that it or they must be supposed to be lost.   This conclusion
will recall the similar problem raised by the *Oenothera* mutants
(Chap. V); and unsatisfactory as it may be to have recourse
to such hypotheses we must remember the possibility that as a
consequence of hybridisation, subsequent segregation and recombination of factors, species may have thus actually, as we
may say, exploded, and left nothing but a polymorphic group of
miscellaneous types to represent them in posterity.   If this way
of regarding the phenomena be a true one, the sterility now seen
when some of the group are re-crossed, becomes analogous to
that " reversion on crossing " which we now so well understand
to be a consequence of the recombination of characters separated
at some previous point in the history of descent.   In the partial
sterility of the contemporary hybrid we see this character re-
appearing, formed now as it was on the occasion of the original
cross, by the meeting of complementary factors.

Another case that may be mentioned in this connection is
that of the crosses between various culinary peas (*Pisum sativum*)
and a peculiar form found by Mr. Arthur Sutton growing ostensibly in a wild state in Palestine.   This Palestine Pea is low
growing, rarely reaching 18 inches.   It is in general appearance
like a small and poorly grown field pea.   The stems are thin and
rather hard.   The most obvious differences which distinguish
this from other field peas are the marked serration of the stipules,
and the development of pith in the pods.   Such pith is often
present in the pods of peas more or less, but in the Palestines it
is so strongly developed as almost to form a lomentum.   Curiously enough, though the flowers are purple much as those of
ordinary field peas, there is no coloured spot in the axils.   On
the other hand, the stems have coloured stripes running up
from the axils.   Though this plant differs so little from domesticated peas, all crosses with them either failed, or produced
hybrids quite or almost quite sterile.   This was Mr. Sutton's
experience, and on repeating the experiments with material
kindly given by him I found the same result.[12]

In a large series of crosses some seeds died or gave rise to

[12] I also had a few F₁ seeds given me by Mr. R. H. Lock.

feeble plants. Of the plants which lived, few gave any seed. The seed, however, that was obtained from F_1 plants grew well enough, and the F_2 plants proved, as often in such cases, fertile. In these, indeed, no sign of sterility was noticeable. The experiment is being repeated in various ways, for, as the genetic behaviour of peas is comparatively well known, the subject is an exceptionally favourable one for these investigations.

Such an example shows the confusion produced the moment we attempt to harmonize conceptions of specific difference with results attained by experimental methods. It has been usual to regard the field pea (*P. arvense*) as a species distinct from the edible pea (*P. sativum*). De Candolle and others regard the field pea as derived from a form wild in Italy, but the origin of the edible pea is considered to be unknown. From breeding experiments we find no sterility whatever in the crosses between the various *arvense* and *sativum* types, nor in the crosses made between them and several other peculiar types from various countries; whereas this Palestine Pea, which only differs from a small *arvense* in what might have been thought trivial characters,[13] either fails to cross altogether or gives a sterile product, whatever type be chosen as the other parent.

Examples of this kind have at least the merit that they lead to more precise delimitations of the problem. We are confronted with two distinct alternatives.

1. We may apply the term Species promiscuously to all distinct forms. If we do so it must be clearly understood that we cannot even rule out the several combinations of " presences and absences " represented by the various types whether wild or domesticated. For we may feel perfectly assured that at least all the *arvense* and all the *sativum* types yet subjected to experimental tests are on precisely the same level in this respect. There is no distinction, logical or physiological, to be drawn between them. Some contain more factors, and others contain fewer. In some the re-combinations have been brought about by natural variation or crossing, while the same consequences in the others have resulted from man's interference.

[13] In a paper about to appear in *Jour. Linn. Soc.* Mr. A. W. Sutton identifies this Palestine pea as *Pisum humile* of Boissier and Noé.

2. We may follow the conventions of systematists and distinguish the outstanding or conspicuous forms such as *arvense*, *quadratum*, *sativum* and perhaps a few more as species, and leave the rest unheeded. If this course is followed it must be clearly understood and permitted as a piece of pure pragmatism, deliberately adopted for the convenience of cataloguers and collectors, without regard to any natural fact or system whatsoever.

But while following either the one plan or the other we shall be still awaiting the answer, which only genetic experiment can provide, to the question whether among the various types there are some which differ from the rest in a peculiar way: whether by having groups of characters linked together in especially durable combinations, or by possessing ingredients which cause greater or less disturbance in the processes of cell-division, and especially in the processes of gametic maturation, when they are united by fertilisation with complementary ingredients.

Before any but the vaguest ideas regarding the nature and significance of inter-specific sterility can be formed, a vast amount of detailed work must be done. Sterility as a result of crossing, as well as that which is alleged sometimes to arise in consequence of changed conditions, is at best a negative characteristic, and there are endless opportunities for mistake and misinterpretation in studying features of this kind. No one, I suppose, would now feel any great confidence in most of the data which from time to time are resuscitated for the purpose of such discussions. Even the best collections of evidence, such as those given by Darwin in *Forms of Flowers*, cannot be regarded as critical when judged by present-day standards. Nothing short of the most familiar acquaintance with the habitual behaviour of individuals, and of strains kept under constant scrutiny for several years would enable the experimenter to form reliable judgments as to the value to be attached to observations of this class.

The admission must, however, be faced that nothing in recent work materially tends to diminish the surprise which has always been felt at the absence of sterility in the crosses between co-derivatives. We should expect such groups of forms to behave like the *Erophila* types, and frequently to produce sterile products on crossing. Whatever be the explanation, the fact remains that such evidence is wanting almost completely. In spite of all that we know of variability nothing readily comparable with the power to produce a sterile hybrid on crossing with a near ally, has yet been observed spontaneously arising, though that characteristic of specificity is one of the most widely distributed in nature. It may be that the lacuna in our evidence is due merely to want of attention to this special aspect of genetic inquiry, and on the whole that is the most acceptable view which can be proposed. But seeing that naturalists are more and more driven to believe the domesticated animals and plants to be poly-phyletic in origin—the descendants, that is to say, of several wild forms—the difficulty is proportionately greater than it was formerly, when variation spontaneously occurring was regarded as a sufficient account of their diversity.

CONCLUDING REMARKS.

The many converging lines of evidence point so clearly to the central fact of the origin of the forms of life by an evolutionary process that we are compelled to accept this deduction, but as to almost all the essential features, whether of cause or mode, by which specific diversity has become what we perceive it to be, we have to confess an ignorance nearly total. The transformation of masses of population by imperceptible steps guided by selection, is, as most of us now see, so inapplicable to the facts, whether of variation or of specificity, that we can only marvel both at the want of penetration displayed by the advocates of such a proposition, and at the forensic skill by which it was made to appear acceptable even for a time.

In place of this doctrine we have little teaching of a positive kind to offer. We have direct perception that new forms of life may arise sporadically, and that they differ from their progenitors quite sufficiently to pass for species. By the success and maintenance of such sporadically arising forms, moreover, there is no reasonable doubt that innumerable strains, whether in isolation or in community with their co-derivatives, have as a fact arisen, which now pass in the lists of systematists as species. For an excellent account of typical illustrations I would refer the reader to the book lately published by R. E. Lloyd[14] on the rat-population of India. The observations there recorded are typical of the state of things disclosed whenever the variations of large numbers of individuals are closely investigated, whether in domestication or in natural conditions.

Guided by such clues we may get a good way into the problem. We see the origin of colourable species in abundance. Then, however, doubt arises whether though these new forms are as good species as many which are accepted as such by even cautious systematists, there may not be a stricter physiological sense in which the term species can be consistently used, which would exclude the whole mass of these *petites espèces*.

If further we find that we have, with certain somewhat doubtful exceptions, never seen the contemporary origin of a dominant factor, or of inter-racial sterility between indubitable co-derivatives, it needs no elaboration of argument to show that the root of the matter has not been reached.

Examination of the inter-relations of unquestionably distinct species nearly allied, such as the two common species of *Lychnis*, leads to the same disquieting conclusion, and the best suggestion we can make as to their origin is that *conceivably* they may have arisen as two re-combinations of factors brought together by the crossing of parent species, one or both of which must be supposed to be lost.

All this is, as need hardly be said, an unsatisfying conclusion. To those permanently engaged in systematics it may well bring despair. The best course for them is once for all to recognise that whether or no specific distinction may prove hereafter to have any actual physiological meaning, it is impossible for the systematist with the means at his disposal to form a judgment of value in any given case. Their business is purely that of the cataloguer, and beyond that they cannot go. They will serve science best by giving names freely and by describing everything

[14] Lloyd, R. E., *The Growth of Groups in the Animal Kingdom*, London, 1912.

to which their successors may possibly want to refer, and generally by subdividing their material into as many species as they can induce any responsible society or journal to publish. Between Jordan with his 200 odd species for *Erophila*, and Grenier and Godron with one, there is no hesitation possible. Jordan's view, as he again and again declares with vehemence, is at least a view of natural facts, whereas the collective species is a mere abstraction, convenient indeed for librarians and beginners, but an insidious misrepresentation of natural truth, perhaps more than any other the source of the plausible fallacies regarding evolution that have so long obstructed progress.

Nevertheless though we have been compelled to retreat from the speculative position to which scientific opinion had rashly advanced, the prospect of permanent progress is greatly better than it was. With the development of genetic research clear conceptions have at length been formed of the kind of knowledge required and of the methods by which it is to be attained. If we no longer see how varieties give rise to species, we may feel confident that a minute study of genetic physiology of varieties and species is the necessary beginning of any critical perception of their inter-relations. It is little more than a century since no valid distinction between a mechanical mixture and a chemical combination could be perceived, and in regard to the forms of life we may well be in a somewhat similar confusion.

As yet the genetic behaviour of animals and plants has only been sampled. When the work has been done on a scale so large as to provide generalisations, we may be in a position to declare whether specific difference is or is not a physiological reality.

INDEX OF SUBJECTS

INDEX OF PERSONS

PRESS OF
THE NEW ERA PRINTING COMPANY
LANCASTER, PA.

Printed in the United States
By Bookmasters